Apache Spark Graph Processing

Build, process, and analyze large-scale graphs
with Spark

Rindra Ramamonjison

BIRMINGHAM - MUMBAI

Apache Spark Graph Processing

First published: September 2015

Production reference: 1040915

Published by Packt Publishing Ltd.
Livery Place
35 Livery Street
Birmingham B3 2PB, UK.

ISBN 978-1-78439-180-5

www.packtpub.com

Credits

Author
Rindra Ramamonjison

Reviewer
Thomas W. Dinsmore
Ryan Mccune
Francoise Provencher

Commissioning Editor
Amit Ghodke

Acquisition Editor
Larissa Pinto

Content Development Editor
Dharmesh Parmar

Technical Editor
Prajakta Mhatre

Copy Editor
Yesha Gangani

Project Coordinator
Nikhil Nair

Proofreader
Safis Editing

Indexer
Tejal Soni

Production Coordinator
Aparna Bhagat

Cover Work
Aparna Bhagat

Foreword

Apache Spark is one of the most compelling technologies in the big data space and for good reason. It allows data scientists and data engineers alike to work in their language of choice (Java, Scala, Python, SQL, and R as of this writing) to make sense of their data. As ReynoldXin noted, Apache Spark is the Swiss Army Knife of big data analytics tools. It allows you to use one tool to do many things from real-time streaming to advanced analytics. And in no small part, the versatility and power of GraphX has helped Spark propel forward.

Apache Spark Graph Processing follows Rindra's journey into solving complex analytics problems. As a PhD graduate in electrical engineering from the University of British Columbia, he focused on applying learning and optimization algorithms to achieve energy-efficient wireless networks. As he dove further into these problems, he realized the ease of which he could solve graph-processing problems by using Apache Spark GraphX. With a tutorial style and hands-on projects with interesting datasets, this book is a reflection of his path from getting started with Apache Spark GraphX to iterative graph parallel processing to learning graph structures.

This book is a great jump-start into GraphX, a practical guide for large-scale graph processing, and a testament to the author's enthusiasm for the Spark community (and the community as a whole).

Denny Lee

Technology Evangelist, Databricks

Advisor, WearHacks

About the Author

Rindra Ramamonjison is a fourth year PhD student of electrical engineering at the University of British Columbia, Vancouver. He received his master's degree from Tokyo Institute of Technology. He has played various roles in many engineering companies, within telecom and finance industries. His primary research interests are machine learning, optimization, graph processing, and statistical signal processing. Rindra is also the co-organizer of the Vancouver Spark Meetup.

About the Reviewer

Thomas W. Dinsmore is a consultant and author with more than 30 years of service to enterprises around the world. He is an expert in business analytics, and has working experience with the leading analytic tools, languages, and databases. In his practice, Thomas helps organizations streamline analytics for improved performance and time to value.

Previously, Thomas served with The Boston Consulting Group, IBM, PriceWaterhouseCoopers and SAS, as well as several startups.

Thomas coauthored *Modern Analytics Methodologies and Advanced Analytics Methodologies*, published in 2014 by FT Press. He is currently under contract to publish a book on disruptive technologies in business analytics, scheduled for publication in Q2 2016.

I would like to thank the entire editorial and production team at Packt Publishing, who work tirelessly to bring quality books to the public.

www.PacktPub.com

Support files, eBooks, discount offers, and more

For support files and downloads related to your book, please visit www.PacktPub.com.

Did you know that Packt offers eBook versions of every book published, with PDF and ePub files available? You can upgrade to the eBook version at www.PacktPub.com and as a print book customer, you are entitled to a discount on the eBook copy. Get in touch with us at service@packtpub.com for more details.

At www.PacktPub.com, you can also read a collection of free technical articles, sign up for a range of free newsletters and receive exclusive discounts and offers on Packt books and eBooks.

https://www2.packtpub.com/books/subscription/packtlib

Do you need instant solutions to your IT questions? PacktLib is Packt's online digital book library. Here, you can search, access, and read Packt's entire library of books.

Why subscribe?
- Fully searchable across every book published by Packt
- Copy and paste, print, and bookmark content
- On demand and accessible via a web browser

Free access for Packt account holders

If you have an account with Packt at www.PacktPub.com, you can use this to access PacktLib today and view 9 entirely free books. Simply use your login credentials for immediate access.

Table of Contents

Preface

This book is intended to present the GraphX library for Apache Spark and to teach the fundamental techniques and recipes to process graph data at scale. It is intended to be a self-study step-by-step guide for anyone new to Spark with an interest in or need for large-scale graph processing.

Distinctive features

The focus of this book is on large-scale graph processing with Apache Spark. The book teaches a variety of graph processing abstractions and algorithms and provides concise and sufficient information about them. You can confidently learn all of it and put it to use in different applications.

- **Step-by-step guide**: Each chapter teaches important techniques for every stage of the pipeline, from loading and transforming graph data to implementing graph-parallel operations and machine learning algorithms.

- **Hands-on approach**: We show how each technique works using the Scala REPL with simple examples and by building standalone Spark applications.

- **Detailed code**: All the Scala code in the book is available for download from the book webpage of Packt Publishing.

- **Real-world examples**: We apply these techniques on open datasets collected from a broad variety of applications ranging from social networks to food science and sports analytics.

What this book covers

This book consists of seven chapters. The first three chapters help you to get started quickly with Spark and GraphX. Then, the next two chapters teach the core techniques and abstractions to manipulate and aggregate graph data. Finally, the last two chapters of this book cover more advanced topics such as graph clustering, implementing graph-parallel iterative algorithms with Pregel, and learning methods from graph data.

Chapter 1, *Getting Started with Spark and GraphX*, begins with an introduction to the Spark system, its libraries, and the Scala Build Tool. It explains how to install and leverage Spark on the command line and in a standalone Scala program.

Chapter 2, *Building and Exploring Graphs*, presents the methods for building Spark graphs using illustrative network datasets.

Chapter 3, *Graph Analysis and Visualization*, walks you through the process of exploring, visualizing, and analyzing different network characteristics.

Chapter 4, *Transforming and Shaping Up Graphs to Your Needs*, teaches you how to transform raw datasets into a usable form that is appropriate for later analysis.

Chapter 5, *Creating Custom Graph Aggregation Operators*, teaches you how to create custom graph operations that are tailored to your specific needs with efficiency in mind, using the powerful message-passing aggregation operator in Spark.

Chapter 6, *Iterative Graph-Parallel Processing with Pregel*, explains the inner workings of the Pregel computational model and describes some use cases.

Chapter 7, *Learning Graph Structures*, introduces graph clustering, which is useful for detecting communities in graphs and applies it to a social music database.

What you need for this book

To learn effectively from this book, it is helpful to have a beginner-level programming experience with Scala. However, intermediate functional constructs or Scala-specific syntax are highlighted and explained as they appear in the book. Prior experience with Spark's core API or with the MapReduce framework is beneficial but not required.

It is also beneficial to follow along with the examples, using a Windows or Unix computer with a Java Development Kit environment. More details on the system requirements are described in the first chapter.

Who this book is for

This book is for data engineers, software developers, and data scientists who need to process graph data at a large scale. This book is intended to be self-contained and does not assume any prior experience with Spark. However, its focus is entirely on graph processing.

Conventions

In this book, you will find a number of text styles that distinguish between different kinds of information. Here are some examples of these styles and an explanation of their meaning.

Code words in text, database table names, folder names, filenames, file extensions, pathnames, dummy URLs, and user input are shown as follows: When we wish to run a Scala code in the Spark shell, it will start with `scala >` prompt.

A block of code is set as follows:

```
val greeting = sc.parallelize("Hello Spark".toList)
greeting.collect foreach {
  c => println(c)
}
```

Any command-line input or output is written as follows:

```
> tar -xf spark-1.4.1-bin-hadoop2.6.tgz
```

New terms and **important words** are shown in bold.

 Warnings or important notes appear in a box like this.

 Tips and tricks appear like this.

Reader feedback

Feedback from our readers is always welcome. Let us know what you think about this book—what you liked or disliked. Reader feedback is important for us as it helps us develop titles that you will really get the most out of.

To send us general feedback, simply e-mail feedback@packtpub.com, and mention the book's title in the subject of your message.

If there is a topic that you have expertise in and you are interested in either writing or contributing to a book, see our author guide at www.packtpub.com/authors.

Customer support

Now that you are the proud owner of a Packt book, we have a number of things to help you to get the most from your purchase.

Downloading the example code

You can download the example code files from your account at http://www.packtpub.com for all the Packt Publishing books you have purchased. If you purchased this book elsewhere, you can visit http://www.packtpub.com/support and register to have the files e-mailed directly to you.

Errata

Although we have taken every care to ensure the accuracy of our content, mistakes do happen. If you find a mistake in one of our books—maybe a mistake in the text or the code—we would be grateful if you could report this to us. By doing so, you can save other readers from frustration and help us improve subsequent versions of this book. If you find any errata, please report them by visiting http://www.packtpub.com/submit-errata, selecting your book, clicking on the **Errata Submission Form** link, and entering the details of your errata. Once your errata are verified, your submission will be accepted and the errata will be uploaded to our website or added to any list of existing errata under the Errata section of that title.

To view the previously submitted errata, go to https://www.packtpub.com/books/content/support and enter the name of the book in the search field. The required information will appear under the **Errata** section.

Piracy

Piracy of copyrighted material on the Internet is an ongoing problem across all media. At Packt, we take the protection of our copyright and licenses very seriously. If you come across any illegal copies of our works in any form on the Internet, please provide us with the location address or website name immediately so that we can pursue a remedy.

Please contact us at copyright@packtpub.com with a link to the suspected pirated material.

We appreciate your help in protecting our authors and our ability to bring you valuable content.

Questions

If you have a problem with any aspect of this book, you can contact us at questions@packtpub.com, and we will do our best to address the problem.

1
Getting Started with Spark and GraphX

Apache Spark is a cluster-computing platform for the processing of large distributed datasets. Data processing in Spark is both fast and easy, thanks to its optimized parallel computation engine and its flexible and unified API. The core abstraction in Spark is based on the concept of **Resilient Distributed Dataset (RDD)**. By extending the MapReduce framework, Spark's Core API makes analytics jobs easier to write. On top of the Core API, Spark offers an integrated set of high-level libraries that can be used for specialized tasks such as graph processing or machine learning. In particular, GraphX is the library to perform graph-parallel processing in Spark.

This chapter will introduce you to Spark and GraphX by building a social network and exploring the links between people in the network. In addition, you will learn to use the **Scala Build Tool (SBT)** to build and run a Spark program. By the end of this chapter, you will know how to:

- Install Spark successfully on your computer
- Experiment with the Spark shell and review Spark's data abstractions
- Create a graph and explore the links using base RDD and graph operations
- Build and submit a standalone Spark application with SBT

Downloading and installing Spark 1.4.1

In the following section, we will go through the Spark installation process in detail. Spark is built on Scala and runs on the **Java Virtual Machine (JVM)**. Before installing Spark, you should first have **Java Development Kit 7 (JDK)** installed on your computer.

Make sure you install JDK instead of **Java Runtime Environment (JRE)**. You can download it from `http://www.oracle.com/technetwork/java/javase/downloads/jdk7-downloads-1880260.html`.

Next, download the latest release of Spark from the project website `https://spark.apache.org/downloads.html`. Perform the following three steps to get Spark installed on your computer:

1. Select the package type: **Pre-built for Hadoop 2.6 and later** and then **Direct Download**. Make sure you choose a prebuilt version for Hadoop instead of the source code.

2. Download the compressed TAR file called `spark-1.4.1-bin-hadoop2.6.tgz` and place it into a directory on your computer.

3. Open the terminal and change to the previous directory. Using the following commands, extract the TAR file, rename the Spark root folder to `spark-1.4.1`, and then list the installed files and subdirectories:

```
tar -xf spark-1.4.1-bin-hadoop2.6.tgz
  mv spark-1.4.1-bin-hadoop2.6 spark-1.4.1
  cd spark-1.4.1
  ls
```

That's it! You now have Spark and its libraries installed on your computer. Note the following files and directories in the `spark-1.4.1` home folder:

*   `core`: This directory contains the source code for the core components and API of Spark

*   `bin`: This directory contains the executable files that are used to submit and deploy Spark applications or also to interact with Spark in a **Spark shell**

*   `graphx`, `mllib`, `sql`, and `streaming`: These are Spark libraries that provide a unified interface to do different types of data processing, namely graph processing, machine learning, queries, and stream processing

*   `examples`: This directory contains demos and examples of Spark applications

It is often convenient to create shortcuts to the Spark home folder and Spark example folders. In Linux or Mac, open or create the `~/.bash_profile` file in your home folder and insert the following lines:

```
export SPARKHOME="/[Where you put Spark]/spark-1.4.1/"
export SPARKSCALAEX="ls ../spark-
1.4.1/examples/src/main/scala/org/apache/spark/examples/"
```

Then, execute the following command for the previous shortcuts to take effect:

```
source ~/.bash_profile
```

As a result, you can quickly access these folders in the terminal or Spark shell. For example, the example named LiveJournalPageRank.scala can be accessed with:

```
$SPARKSCALAEX/graphx/LiveJournalPageRank.scala
```

Experimenting with the Spark shell

The best way to learn Spark is through the Spark shell. There are two different shells for Scala and Python. But since the GraphX library is the most complete in Scala at the time this book was written, we are going to use the spark-shell, that is, the Scala shell. Let's launch the Spark shell inside the $SPARKHOME/bin from the command line:

```
$SPARKHOME/bin/spark-shell
```

If you set the current directory (cd) to $SPARKHOME, you can simply launch the shell with:

```
cd $SPARKHOME
./bin/spark-shell
```

If you happen to get an error saying something like: Failed to find Spark assembly in spark-1.4.1/assembly/target/ scala-2.10. You need to build Spark before running this program, then it means that you have downloaded the Spark source code instead of a prebuilt version of Spark. In that case, go back to the project website and choose a prebuilt version of Spark.

If you were successful in launching the Spark shell, you should see the welcome message like this:

```
Welcome to


    ____              __
   / __/__  ___ _____/ /__
  _\ \/ _ \/ _ `/ __/  '_/
 /___/ .__/\_,_/_/ /_/\_\   version 1.4.1
    /_/

Using Scala version 2.10.4 (Java HotSpot(TM) 64-Bit Server VM, Java)
```

For a sanity check, you can type in some Scala expressions or declarations and have them evaluated. Let's type some commands into the shell now:

```
scala> sc
res1: org.apache.spark.SparkContext = org.apache.spark.
SparkContext@52e52233
scala> val myRDD = sc.parallelize(List(1,2,3,4,5))
myRDD: org.apache.spark.rdd.RDD[Int] = ParallelCollectionRDD[0] at
parallelize at <console>:12
scala> sc.textFile("README.md").filter(line => line contains "Spark").
count()
res2: Long = 21
```

Here is what you can tell about the preceding code. First, we displayed the Spark context defined by the variable `sc`, which is automatically created when you launch the Spark shell. The Spark context is the point of entry to the Spark API. Second, we created an RDD named `myRDD` that was obtained by calling the `parallelize` function for a list of five numbers. Finally, we loaded the `README.md` file into an RDD, filtered the lines that contain the word `"Spark"`, and finally invoked an action on the filtered RDD to count the number of those lines.

> It is expected that you are already familiar with the basic RDD transformations and actions, such as map, reduce, and filter. If that is not the case, I recommend that you learn them first, perhaps by reading the programming guide at `https://spark.apache.org/docs/latest/programming-guide.html` or an introductory book such as *Fast Data Processing with Spark* by Packt Publishing and *Learning Spark* by O'Reilly Media.

Don't panic if you did not fully grasp the mechanisms behind RDDs. The following refresher, however, helps you to remember the important points. RDD is the core data abstraction in Spark to represent a distributed collection of large datasets that can be partitioned and processed in parallel across a cluster of machines. The Spark API provides a uniform set of operations to transform and reduce the data within an RDD. On top of these abstractions and operations, the GraphX library also offers a flexible API that enables us to create graphs and operate on them easily.

Perhaps, when you ran the preceding commands in the Spark shell, you were overwhelmed by the long list of logging statements that start with INFO. There is a way to reduce the amount of information that Spark outputs in the shell.

You can reduce the level of verbosity of the Spark shell as follows:

- First, go to the $SCALAHOME/conf folder
- Then, create a new file called log4j.properties
- Inside the conf folder, open the template file log4j.properties.template and copy all its content into log4j.properties
- Find and replace the line log4j.rootCategory=INFO, console with either one of these two lines:
  - log4j.rootCategory=WARN, console
  - log4j.rootCategory=ERROR, console
- Finally, restart the Spark shell and you should now see fewer logging messages in the shell outputs

Getting started with GraphX

Now that we have installed Spark and experimented with the Spark shell, let's create our first graph in Spark by writing our code in the shell, and then building upon that code to develop and run a standalone program. We have three learning goals in this section:

1. First, you will learn how to construct and explore graphs using the Spark Core and GraphX API through a concrete example.
2. Second, you will review some important Scala programming features that are important to know when doing graph processing in Spark.
3. Third, you will learn how to develop and run a standalone Spark application.

Building a tiny social network

Let's create a tiny social network and explore the relationships among the different people in the network. Again, the best way to learn Spark is inside the shell. Our workflow is therefore to first experiment in the shell and then migrate our code later into a standalone Spark application. Before launching the shell, make sure to change the current directory to $SPARKHOME.

First, we need to import the GraphX and RDD module, as shown, so that we can invoke its APIs with their shorter names:

```scala
scala> import org.apache.spark.graphx._
scala> import org.apache.spark.rdd.RDD
```

As said previously, `SparkContext` is the main point of entry into a Spark program and it is created automatically in the Spark shell. It also offers useful methods to create RDDs from local collections, to load data from a local or Hadoop file system into RDDs, and to save output data on disks.

Loading the data

In this example, we will work with two CSV files `people.csv` and `links.csv`, which are contained in the directory `$SPARKHOME/data/`. Let's type the following commands to load these files into Spark:

```scala
scala> val people = sc.textFile("./data/people.csv")

people: org.apache.spark.rdd.RDD[String] = ./data/people.csv
MappedRDD[81] at textFile at <console>:33


scala> val links = sc.textFile("./data/links.csv")

links: org.apache.spark.rdd.RDD[String] = ./data/links.csv MappedRDD[83]
at textFile at <console>:33
```

Loading the CSV files just gave us back two RDDs of strings. To create our graph, we need to parse these strings into two suitable collections of vertices and edges.

 It is important that your current directory inside the shell is $SPARKHOME. Otherwise, you get an error later because Spark cannot find the files.

The property graph

Before going further, let's introduce some key definitions and graph abstractions. In Spark, a graph is represented by a **property graph**, which is defined in the `Graph` class as:

```scala
class Graph[VD, ED] {
  val vertices: VertexRDD[VD]
    val edges: EdgeRDD[ED,VD]
}
```

This means that the `Graph` class provides getters to access its vertices and its edges. These are later abstracted by the RDD subclasses `VertexRDD[VD]` and `EdgeRDD[ED, VD]`. Note that `VD` and `ED` here denote some Scala-type parameters of the classes `VertexRDD`, `EdgeRDD`, and `Graph`. These types of parameters can be primitive types, such as `String`, or also user-defined classes, such as the `Person` class, in our example of a social graph. It is important to note that the property graph in Spark is a directed multigraph. It means that the graph is permitted to have multiple edges between any pair of vertices. Moreover, each edge is directed and defines a unidirectional relationship. This is easy to grasp, for instance, in a Twitter graph where a user can follow another one but the converse does not need to be true. To model bidirectional links, such as a Facebook friendship, we need to define two edges between the nodes, and these edges should point in opposite directions. Additional properties about the relationship can be stored as an attribute of the edge.

A property graph is a graph with user-defined objects attached to each vertex and edge. The classes of these objects describe the properties of the graph. This is done in practice by parameterizing the class `Graph`, `VertexRDD`, and `EdgeRDD`. Moreover, each edge of the graph defines a unidirectional relationship but multiple edges can exist between any pair of vertices.

Transforming RDDs to VertexRDD and EdgeRDD

Going back to our example, let's construct the graph in three steps, as follows:

1. We define a case class `Person`, which has `name` and `age` as class parameters. Case classes are very useful when we need to do pattern matching on an object `Person` later on:

   ```
   case class Person(name: String, age: Int)
   ```

2. Next, we are going to parse each line of the CSV texts inside people and links into new objects of type `Person` and `Edge` respectively, and collect the results in `RDD[(VertexId, Person)]` and `RDD[Edge[String]]`:

   ```
   val peopleRDD: RDD[(VertexId, Person)] = people map { line
   =>
     val row = line split ','
     (row(0).toInt, Person(row(1), row(2).toInt))
   }
   ```

```
scala> type Connection = String
scala> val linksRDD: RDD[Edge[Connection]] = links map {line =>
  val row = line split ','
  Edge(row(0).toInt, row(1).toInt, row(2))
}
```

To paste or write code in multiple lines in the shell:
- Type the command :paste
- Paste or write the given code
- Evaluate the code by pressing the keys *Cmd + D* on Mac or *Ctrl + D* in Windows

VertexId is simply a type alias for Long as defined in GraphX. In addition, the Edge class is defined in org.apache.spark.graphx.Edge as:

```
class Edge(srcId: VertexId, dstId: VertexId, attr: ED)
```

The class parameters srcId and dstId are the vertex IDs of the source and destination, which are linked by the edge. In our social network example, the link between two people is unidirectional and its property is described in the attr of type Connection. Note that we defined Connection as a type alias for String. For clarity, it often helps to give a meaningful name to the type parameter of Edge.

3. Now, we can create our social graph and name it tinySocial using the factory method Graph(...):

```
scala> val tinySocial: Graph[Person, Connection] =
Graph(peopleRDD, linksRDD)

tinySocial: org.apache.spark.graphx.Graph[Person,Connection] =
org.apache.spark.graphx.impl.GraphImpl@128cd92a
```

There are two things to note about this constructor. I told you earlier that the member vertices and edges of the graph are instances of VertexRDD[VD] and EdgeRDD[ED,VD]. However, we passed RDD[(VertexId, Person)] and RDD[Edge[Connection]] into the above factory method Graph. How did that work? It worked because VertexRDD[VD] and EdgeRDD[ED,VD] are subclasses of RDD[(VertexId, Person)] and RDD[Edge[Connection]] respectively. In addition, VertexRDD[VD] adds the constraint that VertexID occurs only once. Basically, two people in our social network cannot have the same vertex ID. Furthermore, VertexRDD[VD] and EdgeRDD[ED,VD] provide several other operations to transform vertex and edge attributes. We will see more of these in later chapters.

Introducing graph operations

Finally, we are going to look at the vertices and edges in the network by accessing and collecting them:

```
scala> tinySocial.vertices.collect()

res: Array[(org.apache.spark.graphx.VertexId, Person)] =
Array((4,Person(Dave,25)), (6,Person(Faith,21)), (8,Person(Harvey,47)),
(2,Person(Bob,18)), (1,Person(Alice,20)), (3,Person(Charlie,30)),
(7,Person(George,34)), (9,Person(Ivy,21)), (5,Person(Eve,30)))

scala> tinySocial.edges.collect()

res: Array[org.apache.spark.graphx.Edge[Connection]] =
Array(Edge(1,2,friend), Edge(1,3,sister), Edge(2,4,brother),
Edge(3,2,boss), Edge(4,5,client), Edge(1,9,friend), Edge(6,7,cousin),
Edge(7,9,coworker), Edge(8,9,father))
```

We used the `edges` and `vertices` getters in the `Graph` class and used the RDD action `collect` to put the result into a local collection.

Now, suppose we want to print only the professional connections that are listed in the following `profLinks` list:

```
val profLinks: List[Connection] = List("Coworker", "Boss",
"Employee","Client", "Supplier")
```

A bad way to arrive at the desired result is to filter the edges corresponding to professional connections, then loop through the filtered edges, extract the corresponding vertices' names, and print the connections between the source and destination vertices. We can write that method in the following code:

```
val profNetwork =
tinySocial.edges.filter{ case Edge(_,_,link) =>
profLinks.contains(link)}
for {
  Edge(src, dst, link) <- profNetwork.collect()
  srcName = (peopleRDD.filter{case (id, person) => id == src}
first)._2.name
  dstName = (peopleRDD.filter{case (id, person) => id == dst}
first)._2.name
} println(srcName + " is a " + link + " of " + dstName)

Charlie is a boss of Bob
Dave is a client of Eve
George is a coworker of Ivy
```

There are two problems with the preceding code. First, it could be more concise and expressive. Second, it is not efficient due to the filtering operations inside the for loop.

Luckily, there is a better alternative. The GraphX library provides two different ways to view data: either as a graph or as tables of edges, vertices, and triplets. For each view, the library offers a rich set operations whose implementations are optimized for execution. That means that we can often process a graph using a predefined graph operation or algorithm, easily. For instance, we could simplify the previous code and make it more efficient, as follows:

```
tinySocial.subgraph(profLinks contains _.attr).
    triplets.foreach(t => println(t.srcAttr.name + " is a " +
    t.attr + " of " + t.dstAttr.name))
  Charlie is a boss of Bob
  Dave is a client of Eve
  George is a coworker of Ivy
```

We simply used the `subgraph` operation to filter the professional links. Then, we used the **triplet view** to access the attributes of the edges and vertices simultaneously. In brief, the triplet operator returns an RDD of `EdgeTriplet[Person, Connection]`. Note that `EdgeTriplet` is simply an alias for the parameterized type of 3-tuple `((VertexId, Person), (VertexId, Person), Connection)` that contains all the information about the source node, the destination node, and the edge property.

Building and submitting a standalone application

Let's conclude this chapter by developing and running a standalone Spark application for our social network example.

Writing and configuring a Spark program

Satisfied with our experiment in the shell, let's now write our first Spark program. Open your favorite text editor and create a new file named `simpleGraph.scala` and put it in the folder `$SPARKHOME/exercises/chap1`. A template for a Spark program looks like the following code:

```
import org.apache.spark.SparkContext
import org.apache.spark.SparkContext._
import org.apache.spark.SparkConf
```

```scala
import org.apache.spark.rdd.RDD
import org.apache.spark.graphx._
object SimpleGraphApp {
  def main(args: Array[String]){

    // Configure the program
    val conf = new SparkConf()
          .setAppName("Tiny Social")
          .setMaster("local")
          .set("spark.driver.memory", "2G")
    val sc = new SparkContext(conf)

    // Load some data into RDDs
    val people = sc.textFile("./data/people.csv")
    val links = sc.textFile("./data/links.csv")

    // After that, we use the Spark API as in the shell
    // ...
  }
}
```

You can also see the entire code of our `SimpleGraph.scala` file in the example files, which you can download from the Packt website.

Downloading the example code

You can download the example code files from your account at `http://www.packtpub.com` for all the Packt Publishing books you have purchased. If you purchased this book elsewhere, you can visit `http://www.packtpub.com/support` and register to have the files e-mailed directly to you.

Let's go over this code to understand what is required to create and configure a Spark standalone program in Scala.

As a Scala program, our Spark application should be constructed within a top-level Scala object, which must have a `main` function that has the signature: `def main(args: Array[String]): Unit`. In other words, the main program accepts an array of strings as a parameter and returns nothing. In our example, the top-level object is `SimpleGraphApp`.

At the beginning of `simpleGraph.scala`, we have put the following import statements:

```
import org.apache.spark.SparkContext
import org.apache.spark.SparkContext._
import org.apache.spark.SparkConf
```

The first two lines import the `SparkContext` class as well as some implicit conversions defined in its companion object. It is not very important to know what the implicit conversions are. Just make sure you import both `SparkContext` and `SparContext._`

 When we worked in the Spark shell, `SparkContext` and `SparContext._` were imported automatically for us.

The third line imports `SparkConf`, which is a wrapper class that contains the configuration settings of a Spark application, such as its application name, the memory size of each executor, and the address of the master or cluster manager.

Next, we have imported some RDD and GraphX-specific class constructors and operators with these lines:

```
import org.apache.spark.rdd.RDD
import org.apache.spark.graphx._
```

The underscore after `org.apache.spark.graphx` makes sure that all public APIs in GraphX get imported.

Within `main`, we had to first configure the Spark program. To do this, we created an object called `SparkConf` and set the application settings through a chain of setter methods on the `SparkConf` object. `SparkConf` provides specific setters for some common properties, such as the application name or master. Alternatively, a generic `set` method can be used to set multiple properties together by passing them as a sequence of key-value pairs. The most common configuration parameters are listed in the following table with their default values and usage. The extensive list can be found at `https://spark.apache.org/docs/latest/configuration.html`:

Spark property name	Usage and default value
`spark.app.name`	This is the name of your application. This will appear in the UI and in the log data.

Spark property name	Usage and default value
`spark.master`	This is the cluster manager to connect to, for example, `spark://host:port`, `mesos://host:port`, `yarn`, or `local`.
`spark.executor.memory`	This is the amount of memory to use per executor process, in the same format as JVM memory strings (for example, 512 M, 2 G). The default value is 1 G.
`spark.driver.memory`	When you run Spark locally with `spark.master=local`, your executor becomes the driver and you need to set this parameter instead of `spark.executor.memory`. The default value is 512 M.
`spark.storage.memoryFraction`	This is the fraction of Java heap to use for Spark's memory cache. The default is 0.6.
`spark.serializer`	This is the class used to serialize objects to be sent over the network or to be cached in serialized form. This is the subclass of the default class `org.apache.spark.serializer.JavaSerializer`.

In our example, we initialized the program as follows:

```
val conf = new SparkConf()
      .setAppName("Tiny Social")
      .setMaster("local")
      .set("spark.driver.memory", "2G")
val sc = new SparkContext(conf)
```

Precisely, we set the name of our application to `"Tiny Social"` and the master to be the local machine on which we submit the application. In addition, the driver memory is set to 2 GB. Should we have set the master to be a cluster instead of local, we can specify the memory per executor by setting `spark.executor.memory` instead of `spark.driver.memory`.

In principle, the driver and executor have different roles and, in general, they run on different processes except when we set the master to be local. The driver is the process that compiles our program into tasks, schedules these tasks to one of more executors, and maintains the physical location of every RDD. Each executor is responsible for executing the tasks, and storing and caching RDDs in memory.

It is not mandatory to set the Spark application settings in the `SparkConf` object inside your program. Alternatively, when submitting our application, we could set these parameters as command-line options of the `spark-submit` tool. We will cover that part in detail in the following sections. In this case, we will just create our `SparkContext` object as:

```
val sc = new SparkContext(new SparkConf())
```

After configuring the program, the next task is to load the data that we want to process by calling utility methods such as `sc.textFile` on the `SparkContext` object `sc`:

```
val people = sc.textFile("./data/people.csv")
val links = sc.textFile("./data/links.csv")
```

Finally, the rest of the program consists of the same operations on RDDs and graphs that we have used in the shell.

> To avoid confusion when passing a relative file path to I/O actions
> such as `sc.textFile()`, the convention used in this book is
> that the current directory of the command line is always set to the
> project root folder. For instance, if our Tiny Social app's root folder
> is `$SPARKHOME/exercises/chap1`, then Spark will look for the
> data to be loaded in `$SPARKHOME/exercises/chap1/data`. This
> assumes that we have put the files in that `data` folder.

Building the program with the Scala Build Tool

After writing our entire program, we are going to build it using the **Scala Build Tool (SBT)**. If you do not have SBT installed on your computer yet, you need to install it first. Detailed instructions on how to install SBT are available at `http://www.scala-sbt.org/0.13/tutorial/index.html` for most operating systems. While there are different ways to install SBT, I recommend using a package manager instead of the manual installation. After the installation, execute the following command to append the SBT installation folder to the `PATH` environment variable:

```
$ export PATH=$PATH:/usr/local/bin/sbtl
```

Once we have SBT properly installed, we can use it to build our application with all its dependencies inside a single JAR package file, also called **uber jar**. In fact, when running a Spark application on several worker machines, an error will occur if some machines do not have the right dependency JAR.

By packaging an uber jar with SBT, the application code and its dependencies are all distributed to the workers. Concretely, we need to create a build definition file in which we set the project settings. Moreover, we must specify the dependencies and the resolvers that help SBT find the packages that are needed by our program. A resolver indicates the name and location of the repository that has the required JAR file. Let's create a file called `build.sbt` in the project root folder `$SPARKHOME/ exercises/chap1` and insert the following lines:

```
name := "Simple Project"

version := "1.0"

scalaVersion := "2.10.4"

libraryDependencies ++= Seq(
  "org.apache.spark" %% "spark-core" % "1.4.1",
  "org.apache.spark" %% "spark-graphx" % "1.4.1"
)

resolvers += "Akka Repository" at "http://repo.akka.io/releases/"
```

By convention, the settings are defined by Scala expressions and they need to be delimited by blank lines. Earlier, we set the project name, its version number, the version of Scala, as well as the Spark library dependencies. To build the program, we then enter the command:

```
$ sbt package
```

This will create a JAR file inside `$SPARKHOME/exercises/chap1/target/ scala-2.10/simple-project_2.10-1.0.jar`.

Deploying and running with spark-submit

Finally, we are going to invoke the `spark-submit` script in `$SPARKHOME/bin/` to run the program from the root directory `$SPARKHOME/exercises/chap1` in the terminal:

```
$ ../../bin/spark-submit --class \
"SimpleGraphApp" \
./target/scala-2.10/simple-project_2.10-1.0.jar
Spark assembly has been built with Hive, including Datanucleus jars on
classpath
Charlie is a boss of Bob
Dave is a client of Eve
George is a coworker of Ivy
```

The required options for the `spark-submit` command are the Scala application object name and the JAR file that we previously built with SBT. In case we did not set the master setting when creating the `SparkConf` object, we also would have to specify the `--master` option in `spark-submit`.

> You can list all the available options for the `spark-submit` script with the command:
>
> **`spark-submit --help`**
>
> More details about how to submit a Spark application to a remote cluster are available at `http://spark.apache.org/docs/latest/submitting-applications.html`.

Summary

In this chapter, we took a whirlwind tour of graph processing in Spark. Specifically, we installed the Java Development Kit, a prebuilt version of Spark and the SBT tool. Furthermore, you were introduced to graph abstraction and operations in Spark by creating a social network in the Spark shell and also in a standalone program.

In the next chapter, you will learn more about how to build and explore graphs in Spark.

2

Building and Exploring Graphs

This chapter aims to teach us how to represent various types of networks and complex systems as property graphs in Spark and GraphX. Before we can describe the behavior, and analyze the inner structure of these systems, we first need to map their components to vertices or nodes, and map the interactions between the individual components to edges or links. Building on what we learned in the previous chapter, we will delve into the details on how graphs are stored and represented in GraphX. In addition, this chapter introduces the language of graph theory, and the basic characteristics of graphs. Throughout this chapter, we will use real-world datasets that we will map to the different types of graphs. The examples include e-mail communication networks, food flavor network, and social ego networks. On completing this chapter, you will understand how to:

- Load data and build Spark graphs in many ways
- Use the join operator to mix external data into existing graphs
- Build bipartite graphs and multigraphs
- Explore graphs and compute their basic statistics

Network datasets

In the previous chapter, we constructed a small social network as a toy example. From this chapter onwards, we are going to work with real-world datasets, drawn from various applications. In fact, graphs are used to represent any complex system as it describes the interactions between the components of the system. Despite the diversity in form, size, nature, and granularity of different systems, graph theory provides a common language, and a set of tools, for representing and analyzing complex systems.

In brief, a graph consists of a set of vertices connected by a set of edges. Each edge represents the relationship between a pair of connected vertices. In this book, we will sometimes use the less technical terms network nodes to refer to vertices, and links to refer to edges. Note that Spark supports multigraphs, that is, it is permitted to have multiple edges between any pair of nodes.

Let's get a preview of the networks that we are going to build in this chapter.

The communication network

The first type of communication network that we will encounter is an **email communication graph**. A history of e-mails that are exchanged within an organization can be mapped to a communication graph, so as to understand the informal structure behind the organization. Such graphs can also be used to determine influential people or the hubs of the organization that might not necessarily be the high-ranked ones. The email communication network is a canonical example of a directed graph, as each e-mail links a source node to the destination node. We will use the **Enron Corpus**, which is a database of e-mails generated by 158 employees of the Enron Corporation. It is one of the only mass collections of corporate e-mails that are open to public on the web. The Enron Corpus is particularly interesting, as it captures all the communication that occurred inside the company before the scandal that led to its bankruptcy. The original dataset was released by William Cohen at CMU, which can be downloaded from `https://www.cs.cmu.edu/~./enron/`. A detailed description of the complete dataset was done by Klimmt and Yang, 2004. A cleaner version of the dataset, which we use here, is provided by Leskovec et al., 2009, and can be obtained from `https://snap.stanford.edu/data/email-Enron.html`.

Flavor networks

Another example that we will borrow from the culinary world is the ingredient-compound network, introduced by Ahn et al., 2011. It is a bipartite graph in the sense that the nodes are divided into two disjoint sets: the **ingredient nodes** and the **compound nodes**. Each link connects an ingredient to a compound when the chemical compound is present in the food ingredient. From the ingredient-compound network, it is also possible to create what is called a **flavor network**. Instead of connecting food ingredients to compounds, the flavor network links pairs of ingredients whenever a pair of ingredients shares at least one chemical compound.

We will build the ingredient-compound network in this chapter, and in *Chapter 4, Transforming and Shaping Up Graphs to Your Needs*, we will construct the flavor network from the ingredient-compound network. Analyzing such graphs is fascinating because they help us understand more about food pairing and food culture. The flavor network can also help food scientists or amateur cooks create new recipes. The datasets that we will use consist of ingredient-compound data and the recipes collected from `http://www.epicurious.com/`, `allrecipes.com`, and `http://www.menupan.com/`. The datasets are available at `http://yongyeol.com/2011/12/15/paper-flavor-network.html`.

Social ego networks

The last dataset that we will explore in this chapter is a collection of social ego networks from Google+. The data was collected by (McAuley and Leskovec, 2012) from the users who had manually shared their social circles using the **share circle** feature. The dataset includes the user profiles, their circles, and their ego networks and can be downloaded from Stanford's SNAP project website at `http://snap.stanford.edu/data/egonets-Gplus.html`.

These datasets are *not* provided with the Spark installation. They must first be downloaded from their source websites and copied into the `$SPARKHOME/data` folder. When different sizes of the datasets are available, we chose to use the smaller version of the datasets to quickly demonstrate the concepts taught in this book.

Graph builders

In GraphX, there are four functions for building a property graph. Each of these functions requires that the data from which the graph is constructed should be structured in a specified manner.

The Graph factory method

The first one is the `Graph` factory method that we have already seen in the previous chapter. It is defined in the apply method of the companion object called `Graph`, which is as follows:

```
def apply[VD, ED](
    vertices: RDD[(VertexId, VD)],
```

```
        edges: RDD[Edge[ED]],
        defaultVertexAttr: VD = null)
    : Graph[VD, ED]
```

As we have seen before, this function takes two RDD collections: `RDD[(VertexId, VD)]` and `RDD[Edge[ED]]` as parameters for the vertices and edges respectively, to construct a `Graph[VD, ED]` parameter. The `defaultVertexAttr` attribute is used to assign the default attribute for the vertices that are present in the edge RDD but not in the vertex RDD. The `Graph` factory method is convenient when the RDD collections of edges and vertices are readily available.

edgeListFile

A more common situation is that your original dataset only represents the edges. In this case, GraphX provides the following `GraphLoader.edgeListFile` function that is defined in `GraphLoader`:

```
def edgeListFile(
        sc: SparkContext,
        path: String,
        canonicalOrientation: Boolean = false,
        minEdgePartitions: Int = 1)
    : Graph[Int, Int]
```

It takes as an argument a path to the file that contains a list of edges. Each line of the file represents an edge of the graph with two integers in the form: `sourceID destinationID`. When reading the list, it ignores any comment line starting with #. Then, it constructs a graph from the specified edges with the corresponding vertices.

The `minEdgePartitions` argument is the minimum number of edge partitions to generate. If the adjacency list is partitioned with more blocks than `minEdgePartitions`, then more partitions will be created.

fromEdges

Similar to `GraphLoader.edgeListFile`, the third function named `Graph.fromEdges` enables you to create a graph from an `RDD[Edge[ED]]` collection. Moreover, it automatically creates the vertices using the `VertexID` parameters specified by the edge RDD, as well as the `defaultValue` argument as a default vertex attribute:

```
def fromEdges[VD, ED](
        edges: RDD[Edge[ED]],
        defaultValue: VD)
    : Graph[VD, ED]
```

fromEdgeTuples

The last graph builder function is `Graph.fromEdgeTuples`, which creates a graph from only an RDD of edge tuples, that is, a collection of the `RDD[(VertexId, VertexId)]` type. It assigns the edges the attribute value 1 by default:

```
def fromEdgeTuples[VD](
    rawEdges: RDD[(VertexId, VertexId)],
    defaultValue: VD,
    uniqueEdges: Option[PartitionStrategy] = None)
  : Graph[VD, Int]
```

Building graphs

Let's now open our Spark shell and build three types of graphs: a directed email communication network, a bipartite graph of ingredient-compound connections, and a multigraph using the previous graph builders.

 Unless otherwise stated, we always assume that the Spark shell is launched from the $SPARKHOME directory. It then becomes the current directory for any relative file path used in this book.

Building directed graphs

The first graph that we will build is the Enron email communication network. If you have restarted your Spark shell, you need to again import the GraphX library. First, create a new folder called `data` inside $SPARKHOME and copy the dataset into it. This file contains the adjacency list of the email communications between the employees. Assuming that the current directory is $SPARKHOME, we can pass the file path to the `GraphLoader.edgeListFile` method:

```
scala> import org.apache.spark.graphx._
import org.apache.spark.graphx._


scala> import org.apache.spark.rdd._
import org.apache.spark.rdd._
```

```
scala> val emailGraph = GraphLoader.edgeListFile(sc, "./data/emailEnron.
txt")
```

```
emailGraph: org.apache.spark.graphx.Graph[Int,Int] = org.apache.spark.
graphx.impl.GraphImpl@609db0e
```

Notice that the `GraphLoader.edgeListFile` method always returns a graph object, whose vertex and edge attributes have a type `Int`. Their default values are 1. We can check this by looking at the first five vertices and edges in the graph:

```
scala> emailGraph.vertices.take(5)
```

```
res: Array[(org.apache.spark.graphx.VertexId, Int)] = Array((19021,1),
(28730,1), (23776,1), (31037,1), (34207,1))
```

```
scala> emailGraph.edges.take(5)
```

```
res: Array[org.apache.spark.graphx.Edge[Int]] = Array(Edge(0,1,1),
Edge(1,0,1), Edge(1,2,1), Edge(1,3,1), Edge(1,4,1))
```

The first node `(19021,1)` has the vertex ID `19021` and its attribute is indeed set to 1. Similarly, the first edge `Edge(0,1,1)` captures the communication between the source 0 and destination 1.

In GraphX, all the edges must be directed. To express non-directed or bidirectional graphs, we can link each connected pair in both directions. In our email network, we can verify for instance that the `19021` node has both incoming and outgoing links. First, we collect the destination nodes that node `19021` communicates to:

```
scala> emailGraph.edges.filter(_.srcId == 19021).map(_.dstId).collect()
```

```
res: Array[org.apache.spark.graphx.VertexId] = Array(696, 4232, 6811,
8315, 26007)
```

It turns out that these same nodes are also the source nodes for the incoming edges to `19021`:

```
scala> emailGraph.edges.filter(_.dstId == 19021).map(_.srcId).collect()
```

```
res: Array[org.apache.spark.graphx.VertexId] = Array(696, 4232, 6811,
8315, 26007)
```

Building a bipartite graph

In some applications, it is useful to represent a view of a system as a bipartite graph. A bipartite graph is composed of two sets of nodes. The nodes within the same set cannot be connected but only the pairs belonging to the different sets can be. An example of such a graph is the food ingredient-compound network.

Here, we will work with the files `ingr_info.tsv`, `comp_info.tsv`, and `ingr_comp.tsv`, which should be copied into the `$SPARKHOME/data` folder. The first two files contain the information about the food ingredients and compounds respectively.

Let's have a quick look at the first lines of these two files using the `Source.fromFile` method of `scala.io.Source`. Our only requirement for this method is to simply inspect the beginning of the text files:

```
scala> import scala.io.Source
import scala.io.Source


scala> Source.fromFile("./data/ingr_info.tsv").getLines().
      take(7).foreach(println)
# id  ingredient name   category
0   magnolia_tripetala   flower
1   calyptranthes_parriculata   plant
2   chamaecyparis_pisifera_oil   plant derivative
3   mackerel   fish/seafood
4   mimusops_elengi_flower   flower
5   hyssop   herb


scala> Source.fromFile("./data/comp_info.tsv").getLines().
take(7).foreach(println)
# id   Compound name   CAS number
0   jasmone   488-10-8
1   5-methylhexanoic_acid   628-46-6
2   l-glutamine   56-85-9
3   1-methyl-3-methoxy-4-isopropylbenzene   1076-56-8
4   methyl-3-phenylpropionate   103-25-3
5   3-mercapto-2-methylpentan-1-ol_(racemic)   227456-27-1
```

The third file contains the adjacency list between the ingredients and the compounds:

```
scala> Source.fromFile("./data/ingr_comp.tsv").getLines().
take(7).foreach(println)
# ingredient id   compound id
1392   906
1259   861
```

1079	673
22	906
103	906
1005	906

In practice, the datasets from which we build the graphs will not come in a form that the graph builders in Spark expect them to be in. For example, in the food network example, we have two problems with the datasets. First, we cannot simply create a graph from the adjacency list because the indices of the ingredients and compounds both start at zero and overlap with each other. Therefore, there is no way to distinguish the two nodes if they happen to have the same vertex ID. Second, we have two kinds of nodes--ingredients and compounds:

> In order to create a bipartite graph, we first need to create the case classes named `Ingredient` and `Compound`, and use Scala's inheritance so that these two classes are the children of a `FNNode` class.

```scala
scala> class FNNode(val name: String)
defined class FNNode
```

```scala
scala> case class Ingredient(override val name: String, category: String)
extends FNNode(name)
defined class Ingredient
```

```scala
scala> case class Compound(override val name: String, cas: String)
extends FNNode(name)
defined class Compound
```

After this, we need to load all the `Compound` and `Ingredient` objects into an `RDD[FNNode]` collection. This part requires some data wrangling:

```scala
val ingredients: RDD[(VertexId, FNNode)] =
sc.textFile("./data/ingr_info.tsv").
    filter(! _.startsWith("#")).
    map {line =>
        val row = line split '\t'
        (row(0).toInt, Ingredient(row(1), row(2)))
    }
ingredients:
org.apache.spark.rdd.RDD[(org.apache.spark.graphx.VertexId,
FNNode)] = MappedRDD[32] at map at <console>:26
```

In the preceding code, we first loaded the text in comp_info.tsv into an RDD of String, and filtered out the comment lines starting with #. Then, we parsed the tab-delimited lines into RDD of Ingredient vertices. Now, let's do a similar thing with comp_info.tsv and create an RDD of Compound vertices:

```
val compounds: RDD[(VertexId, FNNode)] =
sc.textFile("./data/comp_info.tsv").
        filter(! _.startsWith("#")).
        map {line =>
                val row = line split '\t'
                (10000L + row(0).toInt, Compound(row(1), row(2)))
            }
compounds:
org.apache.spark.rdd.RDD[(org.apache.spark.graphx.VertexId,
FNNode)] = MappedRDD[28] at map at <console>:26
```

However, there is a critical thing that we did earlier. Since the index of each node should be unique, we had to shift the range of the compound indices by 10000L, so that there is no index that refers to an ingredient and a compound at the same time.

Next, we create an RDD[Edge[Int]] collection from the dataset named ingr_comp.tsv:

```
val links: RDD[Edge[Int]] =
    sc.textFile("./data/ingr_comp.tsv").
        filter(! _.startsWith("#")).
        map {line =>
            val row = line split '\t'
            Edge(row(0).toInt, 10000L + row(1).toInt, 1)
        }
```

When parsing the lines of the adjacency list in ingr_comp.tsv, we also shift the indices of compounds by 10000L. This quick fix works perfectly because we knew in advance, from the dataset description, how many ingredients and compounds there were in the dataset. Be more careful with real messy datasets! Next, as the links between ingredients and compounds do not contain any weight or meaningful attributes, we just parameterized the Edge class with the Int type, and set a default value of 1 for the attribute of each link.

Finally, we concatenate the two sets of nodes into a single RDD, and pass it to the Graph() factory method along with the RDD link:

```
scala> val nodes = ingredients ++ compounds
```

```
nodes: org.apache.spark.rdd.RDD[(org.apache.spark.graphx.VertexId,
FNNode)] = UnionRDD[61] at $plus$plus at <console>:30


scala> val foodNetwork = Graph(nodes, links)

foodNetwork: org.apache.spark.graphx.Graph[FNNode,Int] = org.apache.
spark.graphx.impl.GraphImpl@635933c1
```

So, let's explore the ingredient-compound graph:

```
scala> def showTriplet(t: EdgeTriplet[FNNode,Int]): String = "The
ingredient " ++ t.srcAttr.name ++ " contains " ++ t.dstAttr.name

showTriplet: (t: EdgeTriplet[FNNode,Int])String


scala> foodNetwork.triplets.take(5).

    foreach(showTriplet _ andThen println _)

The ingredient calyptranthes_parriculata contains citral_(neral)

The ingredient chamaecyparis_pisifera_oil contains undecanoic_acid

The ingredient hyssop contains myrtenyl_acetate

The ingredient hyssop contains 4-(2,6,6-trimethyl-cyclohexa-1,3-dienyl)
but-2-en-4-one

The ingredient buchu contains menthol
```

First, we defined a helper function called `showTriplet` that returns a `String`
description of an ingredient-compound triplet. Then, we took the first five triplets
and printed them out on the console. In the preceding example, we used Scala's
function composition in the `showTriplet _ andThen println _` argument and
it passed to the `foreach` method.

Building a weighted social ego network

As a final example, let's build an ego network from the Google+ dataset that we
presented earlier in this chapter. An ego network is a graph representation of one
person's connections. Precisely, it focuses on a single node called the focal node and
only represents the links between that node and its neighbors. Although the entire
dataset from the SNAP website contains the ego networks of 133 Google+ users, we
are only going to build one person's ego network as an illustration. The files that we
are going to work with are placed in `$SPARKHOME/data`.

Their description is given as follows:

- `ego.edges`: These are directed edges in the ego network. The `ego` node does not appear in this list, but it is assumed that it follows every node ID that appears in the file.

- `ego.feat` : This features for each of the nodes that appear in the edge file.

- `ego.featnames`: This is the name of each of the feature dimensions. The feature is 1 if the user has this property in their profile, and 0 otherwise.

First, let's import the absolute value function and the `SparseVector` class from the Breeze library, which we will be using:

```
import scala.math.abs
import breeze.linalg.SparseVector
```

Then, let's also define a type synonym called `Feature` for `SparseVector[Int]`:

```
type Feature = breeze.linalg.SparseVector[Int]
```

Using the following code, we can read the features inside the `ego.feat` file and collect them in a map whose keys and values are of the `Long` and `Feature` types, respectively:

```
val featureMap: Map[Long, Feature] =
  Source.fromFile("./data/ego.feat").
      getLines().
      map{line =>
      val row = line split ' '
      val key = abs(row.head.hashCode.toLong)
      val feat = SparseVector(row.tail.map(_.toInt))
      (key, feat)
      }.toMap
```

Let's step back and take a quick look inside the ego.feat file to understand what the preceding chain of RDD transformations is doing, and why it is needed. Each line in ego.feat has the following form:

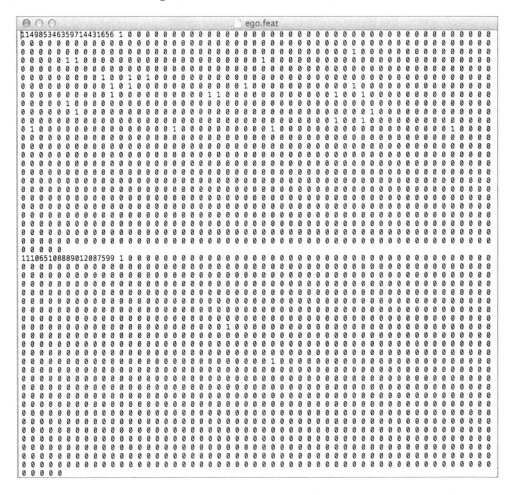

The first number in each line corresponds to a node's ID in the ego network. The remaining string of 0 and 1 numbers indicate which feature this particular node has. For example, the first 1 after the node's ID corresponds to the gender:1 feature. In fact, each feature is by the design of the description:value form. In practice, we usually have a limited control over the format of the datasets that we are working with. As in this example, there is always some data wrangling that we need to do. First, each vertex in the ego network should have a vertex ID of the Long type. However, the node IDs in the dataset, such as 1149853463597114431656, exceed the permitted range for Long.

Therefore, we have to create new indices for the nodes. Second, we need to parse the string of 0 and 1 in the data to create a feature vector that has a more convenient form.

Luckily, these issues do have easy fixes. To convert the original node ID to a vertex ID, we simply hash the string that corresponds to the node ID, as follows:

```
val key = abs(row.head.hashCode.toLong)
```

Then, we took advantage of the `SparseVector` representation in the Breeze library to efficiently store the feature indices.

Next, we can read the `ego.edges` file to create an `RDD[Edge[Int]]` collection of the links in the ego network. In contrast to our previous graph examples, we model the ego network as a weighted graph. Precisely, the attribute of each link will correspond to the number of common features that each connected pair has. This is done by the following transformations:

```
val edges: RDD[Edge[Int]] =
  sc.textFile("./data/ego.edges").
    map {line =>
      val row = line split ' '
      val srcId = abs(row(0).hashCode.toLong)
      val dstId = abs(row(1).hashCode.toLong)
      val srcFeat = featureMap(srcId)
      val dstFeat = featureMap(dstId)
      val numCommonFeats = srcFeat dot dstFeat
      Edge(srcId, dstId, numCommonFeats)
    }
```

> To find the number of common features between the source and destination nodes, we just used the dot product operation of the `SparseVector` class in Breeze. Again, we also had to compute new vertex IDs using the `hashCode` attribute of the node IDs in the dataset.

Finally, we can now create an ego network using the `Graph.fromEdges` function. This function takes as arguments the `RDD[Edge[Int]]` collection and the default value for the vertices:

```
val egoNetwork: Graph[Int,Int] = Graph.fromEdges(edges, 1)
```

Then, we can check how many of the nodes in the ego node's connections have some features in common with their adjacent nodes:

```scala
scala> egoNetwork.edges.filter(_.attr == 3).count()
res: Long = 1852


scala> egoNetwork.edges.filter(_.attr == 2).count()
res: Long = 9353


scala> egoNetwork.edges.filter(_.attr == 1).count()
res: Long = 107934
```

Computing the degrees of the network nodes

We are now going to explore the three graphs, and introduce an important property of a network node, which is the degree of the node.

The degree of a node represents the number of links it has to other nodes. In a directed graph, we can make a distinction between the incoming degree of a node or an in-degree, which is the number of its incoming links, and its outgoing degree or out-degree, which is the number of nodes that it points to. In the following sections, we will explore the degree distributions of the three example networks.

In-degree and out-degree of the Enron email network

For the Enron email network, we can confirm that there are roughly ten times more links than nodes:

```scala
scala> emailGraph.numEdges
res: Long = 367662


scala> emailGraph.numVertices
res: Long = 36692
```

Indeed, the in-degree and out-degree of the employees are exactly the same in this example as the email graph is bi-directed. This can be confirmed by looking at the average degrees:

```scala
scala> emailGraph.inDegrees.map(_._2).sum / emailGraph.numVertices
res: Double = 10.020222391802028
```

```scala
scala> emailGraph.outDegrees.map(_._2).sum / emailGraph.numVertices
res: Double = 10.020222391802028
```

If we want to find the person that has e-mailed to the largest number of people, we can define and use the following max function:

```scala
def max(a: (VertexId, Int), b: (VertexId, Int)): (VertexId, Int) = {
  if (a._2 > b._2) a else b
}
```

Let's see the output:

```scala
scala> emailGraph.outDegrees.reduce(max)
res: (org.apache.spark.graphx.VertexId, Int) = (5038,1383)
```

This person could be an executive or an employee, acting as a hub to the organization. Similarly, we can define a min function to find people. Now, let's check if there are some isolated groups of employees at Enron using the following code:

```scala
scala> emailGraph.outDegrees.filter(_._2 <= 1).count
res83: Long = 11211
```

It seems that there are many employees who receive e-mails from only one employee — perhaps their bosses or from the human resources department.

Degrees in the bipartite food network

For the bipartite ingredient-compound graph, we can also explore which food has the largest number of compounds, or which compound is the most prevalent in our list of ingredients:

```scala
scala> foodNetwork.outDegrees.reduce(max)
res: (org.apache.spark.graphx.VertexId, Int) = (908,239)
```

```scala
scala> foodNetwork.vertices.filter(_._1 == 908).collect()
```

```
res: Array[(org.apache.spark.graphx.VertexId, FNNode)] =
Array((908,Ingredient(black_tea,plant derivative)))
```

```
scala> foodNetwork.inDegrees.reduce(max)
```

```
res: (org.apache.spark.graphx.VertexId, Int) = (10292,299)
```

```
scala> foodNetwork.vertices.filter(_._1 == 10292).collect()
```

```
res: Array[(org.apache.spark.graphx.VertexId, FNNode)] =
Array((10292,Compound(1-octanol,111-87-5)))
```

The answers to the earlier two questions turn out to be the black tea and the compound `1-octanol`.

Degree histogram of the social ego networks

Similarly, we can compute the degrees of the connections in the ego network. Let's look at the maximum and minimum degrees in the network:

```
scala> egoNetwork.degrees.reduce(max)
```

```
res91: (org.apache.spark.graphx.VertexId, Int) = (1643293729,1084)
```

```
scala> egoNetwork.degrees.reduce(min)
```

```
res92: (org.apache.spark.graphx.VertexId, Int) = (550756674,1)
```

Suppose that we now want to have the histogram data of the degrees. Then, we can write the following code to do just that:

```
egoNetwork.degrees.
  map(t => (t._2,t._1)).
  groupByKey.map(t => (t._1,t._2.size)).
  sortBy(_._1).collect()

res: Array[(Int, Int)] = Array((1,15), (2,19), (3,12), (4,17),
(5,11), (6,19), (7,14), (8,9), (9,8), (10,10), (11,1), (12,9),
(13,6), (14,7), (15,8), (16,6), (17,5), (18,5), (19,7), (20,6),
(21,8), (22,5), (23,8), (24,1), (25,2), (26,5), (27,8), (28,4),
(29,6), (30,7), (31,5), (32,10), (33,6), (34,10), (35,5), (36,9),
(37,7), (38,8), (39,5), (40,4), (41,3), (42,1), (43,3), (44,5),
(45,7), (46,6), (47,3), (48,6), (49,1), (50,9), (51,5),...
```

Summary

In this chapter, we have learned about the different ways to build graphs in Spark by working with concrete examples borrowed from online social networks, food science, and e-mail communications. We have seen that constructing a graph requires some data preparation and wrangling efforts. Nonetheless, GraphX offers various graph builder functions from which we can choose, depending on the graph representation that we need to create, and on the shape of the available datasets. Such usable functionalities are the advantages of GraphX against other similar graph-processing frameworks. Moreover, we looked at some basic statistics and properties of graphs, which are rather useful in characterizing their structure and in understanding their representation.

In the next chapter, we will go deeper into the analysis of graphs, using data visualization tools and new graph-theoretical concepts and algorithms, such as connectedness, triangle counting, and PageRank.

3

Graph Analysis and Visualization

In this chapter, we will learn how to analyze the characteristics of graphs using visualization tools and graph algorithms. For example, we will use some of the algorithms available in GraphX to see how connected a graph is. In addition, we will compute metrics that are commonly used, such as triangle counting and clustering coefficients. Furthermore, we will learn through a concrete example how the PageRank algorithm can be used to rank the importance of the nodes in a network. Along the way, we will introduce new RDD operations that will prove out to be useful here and in later chapters. Finally, this chapter offers practical tips on building Spark applications that rely on the third-party libraries. After doing the activities in this chapter, you will have learned the tools and concepts to:

- Visualize large-scale graph data
- Compute the connected components of a network
- Use the PageRank algorithm to rank the node importance in networks
- Build Spark applications that use third-party libraries using SBT

Network datasets

We will be using the same datasets introduced in *Chapter 2*, *Building and Exploring Graphs*, including the social ego network, email graph, and food-compound network.

The graph visualization

Spark and GraphX do not provide any built-in functionality for data visualization, since their focus is on data processing. However, pictures are worth than thousands of numbers when it comes to data analysis. In the following sections, we will build a Spark application for visualizing and analyzing the connectedness of graphs. We will rely on the third-party library called **GraphStream** for drawing networks, and **BreezeViz** for plotting structural properties of graphs, such as degree distribution. These libraries are not perfect and have limitations but they are relatively stable and simple to use. So, we will use them for exploring the graph examples that are used in this chapter.

 Currently, there is still a lack of graph visualization engines and libraries for drawing large-scale networks, without requiring a huge amount of computing resources. For example, the popular network analysis software SNAP currently relies on the GraphViz engine to draw networks, but it can only draw small- to medium-sized networks. Gephi is another tool for doing interactive network visualization. Although it has nice features, such as a multilevel layout and a built-in 3D rendering engine, Gephi still requires a high CPU and memory requirements. For drawing standards plots, the new project Apache Zeppelin offers a web-based notebook for interactive data analysis and visualization. It also provides a built-in integration with Spark. Visit the official website for more information.

Installing the GraphStream and BreezeViz libraries

Let's get going by installing the third-party libraries and their dependencies in the $SPARKHOME /lib folder. GraphStream is an awesome Java library that enables the visualization of dynamic networks, which can evolve with time. For our purpose, we only need to display static networks so that we only need to download two JAR files called gs-core-1.2.jar and gs-ui-1.2.jar for the core and UI libraries. They can be downloaded from the following repositories:

- https://oss.sonatype.org/content/repositories/releases/org/
  graphstream/gs-core/1.2/

- `https://oss.sonatype.org/content/repositories/releases/org/graphstream/gs-ui/1.2/`

Put these two JAR files in the `lib` folder, within the project root directory. Next, download the `breeze_2.10-0.9.jar` and `breeze-viz_2.10-0.9.jar` libraries from the following repositories:

- `http://repo.spring.io/libs-release-remote/org/scalanlp/breeze_2.10/0.9/`

- `http://repo1.maven.org/maven2/org/scalanlp/breeze-viz_2.10/0.9/`

Since BreezeViz is a Scala library that depends on another Java library called **JfreeChart**, you will also need to install `jcommon-1.0.16.jar` and `jfreechart-1.0.13.jar`. These JAR files can be found in the following repositories:

- `https://repository.jboss.org/nexus/content/repositories/thirdparty-releases/jfree/jcommon/1.0.16/`

- `http://repo1.maven.org/maven2/jfree/jfreechart/1.0.13/`

After you have downloaded all these four JAR files, copy them into the `lib` folder within the project root directory. You are now ready to draw your first graph from the Spark shell.

Visualizing the graph data

Open the terminal, with the current directory set to `$SPARKHOME`. Launch the Spark shell. This time you will need to specify the third-party JAR files with the `--jars` option:

```
$ ./bin/spark-shell --jars \
lib/breeze-viz_2.10-0.9.jar,\
lib/breeze_2.10-0.9.jar,\
lib/gs-core-1.2.jar,\
lib/gs-ui-1.2.jar,\
lib/jcommon-1.0.16.jar,\
lib/jfreechart-1.0.13.jar
```

Alternatively, you can save yourself some typing with the shorter command:

```
$./bin/spark-shell   --jars \
$(find "." -name '*.jar' | xargs echo | tr ' ' ',')
```

Instead of specifying each JAR one at a time, the preceding command loads all the JARs.

As a first example, we will visualize the social ego network that we have seen in the previous chapter. First, we need to import the GraphStream classes with the following:

```scala
scala> import org.graphstream.graph.{Graph => GraphStream}
import org.graphstream.graph.{Graph=>GraphStream}
scala> import org.graphstream.graph.implementations._
import org.graphstream.graph.implementations._
```

It is important that we rename org.graphstream.graph.Graph to GraphStream, to avoid a namespace collision with the Graph class of GraphX.

Next, load the social ego network data using Graph.fromEdges, as we did in the previous chapter. After that, we will create a SingleGraph object:

```scala
// Create a SingleGraph class for GraphStream visualization
val graph: SingleGraph = new SingleGraph("EgoSocial")
```

The SingleGraph object is a GraphStream abstraction that enables the manipulation and visualization of graph data. Concretely, we can invoke the addNode and addEdge methods of the SingleGraph object to add the network nodes and links. We can also invoke the addAttribute method on either the graph, or each individual edge and node to set their visual attributes. What's cool about the GraphStream API is that it cleanly separates the graph structure and visualization using a CSS-like style sheet to control the way the graph elements are displayed. It is much easier to see this in action. So, let's create a file named stylesheet and put it in a new ./style/ folder. Insert the following lines in the style sheet:

```
node {
    fill-color: #a1d99b;
    size: 20px;
    text-size: 12;
    text-alignment: at-right;
    text-padding: 2;
    text-background-color: #fff7bc;
}
edge {
    shape: cubic-curve;
    fill-color: #dd1c77;
    z-index: 0;
    text-background-mode: rounded-box;
    text-background-color: #fff7bc;
    text-alignment: above;
    text-padding: 2;
}
```

The preceding style sheet describes the visual styles of the graph elements using selectors nodes and edges, and specifying their visual attributes using key-value pairs. In this example, we set the colors and shapes of the nodes, edges, and their text attributes. An exhaustive reference for the style sheet attributes used in GraphStream is available at http://graphstream-project.org/doc/Tutorials/Graph-Visualisation_1.1/.

With the style sheet now ready, we can connect it to the SingleGraph object graph:

```
// Set up the visual attributes for graph visualization
graph.addAttribute("ui.stylesheet","url(file:.//style/stylesheet)")
graph.addAttribute("ui.quality")
graph.addAttribute("ui.antialias")
```

In the last two lines, we simply informed the rendering engine to favor quality instead of speed. Next, we have to reload the graph that we built in the previous chapter. To avoid repetitions, we omit the graph building part. After this, we now load VertexRDD and EdgeRDD of the social network into the GraphStream graph object, with the following code:

```
// Given the egoNetwork, load the graphX vertices into GraphStream
for ((id,_) <- egoNetwork.vertices.collect()) {
val node = graph.addNode(id.toString).asInstanceOf[SingleNode]
}
// Load the graphX edges into GraphStream edges
for (Edge(x,y,_) <- egoNetwork.edges.collect()) {
val edge = graph.addEdge(x.toString ++ y.toString,
x.toString, y.toString,
true).
    asInstanceOf[AbstractEdge]
}
```

To add a node, we simply pass its vertex ID as a string argument. For the edges, we need to pass four arguments to the addEdge method. The first one is a string identifier for each edge. Since this identifier is not available in the original dataset or in the GraphX graph, we had to create one. Well, here the simplest solution was to concatenate the vertex IDs of the nodes that each edge links to.

In the preceding code, we had to use a subtle trick to avoid an interoperability issue between our Scala code and the `GraphStream` Java library. As described in the `org.graphstream.graph.implementations.AbstractGraph` API of `GraphStream`, the `addNode` and `addEdge` methods return the node and edge respectively. However, as `GraphStream` is a third-party Java library, we had to force the return types of `addNode` and `addEdge` using the `asInstanceOf[T]` method with the `T` type being `SingleNode` and `AbstractEdge`, respectively. So what would have happened if we omitted these explicit type conversions? You would get a rather strange exception saying:

```
java.lang.ClassCastException:
org.graphstream.graph.implementations.SingleNode cannot
be cast to scala.runtime.Nothing$
```

Now what? The only thing to do here is to make the social ego network display it. Just call the display method on graph:

```
graph.display()
```

Voila! You now will see the network drawn in a new window, as shown in the following:

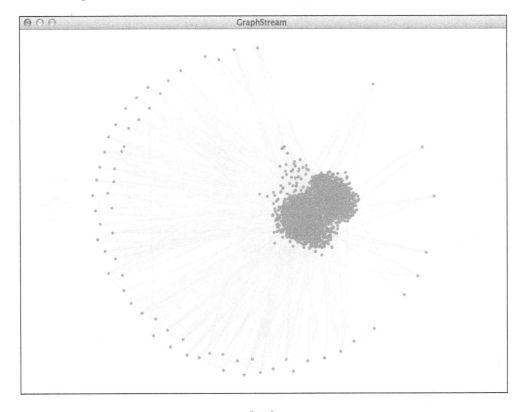

If your graph is not displayed with the colors above, you should check that the stylesheet's file path is correct when setting the graph's attribute called `ui.stylesheet`.

Plotting the degree distribution

As shown by this visualization, each person in the ego network seems to be either isolated or connected to a large group of mutual friends. We can further analyze this fact by plotting the degree distribution of the network. To do this with the help of the Spark shell is as easy as before. Make sure that you first import some classes from JFreeChart and Breeze:

```
import org.jfree.chart.axis.ValueAxis
import breeze.linalg._
import breeze.plot._
```

We will then employ the `degreeHistogram` function that we built in *Chapter 2, Building and Exploring Graphs*. For convenience, its definition is shown as follows:

```
def degreeHistogram(net: Graph[Int, Int]): Array[(Int, Int)] =
    net.degrees.map(t => (t._2,t._1)).
        groupByKey.map(t => (t._1,t._2.size)).
        sortBy(_._1).collect()
```

From the degree histogram, we can obtain the degree distribution, which is the probability distribution of the node degrees over the whole network. For this, we just normalize the node degrees by the total number of nodes, so that the degree probabilities add up to one:

```
val nn = egoNetwork.numVertices
val egoDegreeDistribution = degreeHistogram(egoNetwork).map({case
(d,n) => (d,n.toDouble/nn)})
```

To display the degree distribution, we first create a `Figure` object called `f` and two plot objects called `p1` and `p2`. In the following code, `p1 = f.subplot(2,1,0)` and `p2 = f.subplot(2,1,1)` specify that `f` will have two subplots, and that `p1` is displayed above `p2`. Indeed, the first two arguments of the subplot are the number of rows and columns of the figure, whereas the third argument denotes the subplot index, which starts at 0:

```
val f = Figure()
val p1 = f.subplot(2,1,0)
val x = new DenseVector(egoDegreeDistribution map (_._1.toDouble))
val y = new DenseVector(egoDegreeDistribution map (_._2))
```

```
p1.xlabel = "Degrees"
p1.ylabel = "Distribution"
p1 += plot(x, y)
p1.title = "Degree distribution of social ego network"
val p2 = f.subplot(2,1,1)
val egoDegrees = egoNetwork.degrees.map(_._2).collect()

p1.xlabel = "Degrees"
p1.ylabel = "Histogram of node degrees"
p2 += hist(egoDegrees, 10)
```

This code will then display the degree distribution and degree frequencies of the ego network:

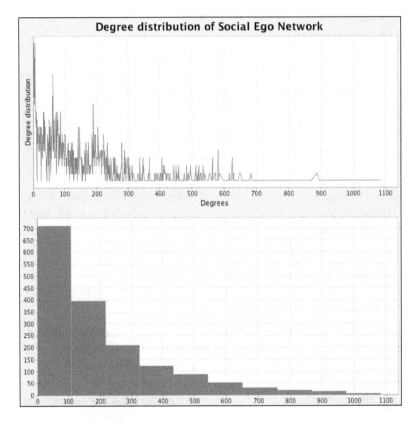

The analysis of network connectedness

Next, we are going to visually explore and analyze the connectedness of the food network. Reload the ingredient and compound datasets using the steps explained in the previous chapter. After you are done, create a `GraphStream` graph object:

```
// Create a SingleGraph class for GraphStream visualization
val graph: SingleGraph = new SingleGraph("FoodNetwork")
```

Then, set the `ui.stylesheet` attribute of the graph. Since the food network is a bipartite graph, it would be nice to display the nodes with two different colors. We do that using a new style sheet. While we are at it, let's also reduce the node size and hide the text attributes:

```
node {
    size: 5px;
    text-mode: hidden;
    z-index: 1;
    fill-mode: dyn-plain;
    fill-color: "#e7298a", "#43a2ca";
}
edge {
    shape: line;
    fill-color: #fee6ce;
    arrow-size: 2px, 1px;
    z-index: 0;
}
```

 The color value in the style sheet is set in hexadecimal using #. You can choose your favorite colors from the awesome **ColorBrewer** palettes available at http://colorbrewer2.org/.

Let's now load the nodes and edges from `foodNetwork` to the `GraphStream` graph again, using the `addNode` and `addEdge` methods. This time, we are going to dynamically set the color of the nodes, depending on whether it is an ingredient or a compound:

```
// Load the graphX vertices into GraphStream nodes
for ((id:VertexId, fnn:FNNode) <- foodNetwork.vertices.collect())
{
val node = graph.addNode(id.toString).asInstanceOf[SingleNode]
node.addAttribute("name", fnn.name)
node.addAttribute("ui.label", fnn.name)
if (fnn.isInstanceOf[Compound])
  node.addAttribute("ui.color",1: java.lang.Double)
```

```
else if(fnn.isInstanceOf[Compound])
    node.addAttribute("ui.color",0: java.lang.Double)
}
```

> You may ask yourself why we used `isInstanceOf[T]` to determine the type of the nodes when loading the nodes with `addNode`. Why did we not use Scala's awesome pattern matching feature? We could have used it in a standalone Spark program, but it is not currently possible to pattern match on case classes in the Spark shell. So, that is why we used `isInstanceOf[T]`.

Loading the nodes of the food network is almost the same as for the social ego network. The only difference is setting different colors for the nodes. In a similar fashion, load the edges into the `GraphStream` graph object:

```
// Load the graphX edges into GraphStream edges
for (Edge(x,y,_) <- foodNetwork.edges.collect()) {
  val edge = graph.addEdge(x.toString ++ y.toString, \
      x.toString, y.toString,
      true).
asInstanceOf[AbstractEdge]
}
```

To visualize the food network, call `graph.display()`. You will get something like this:

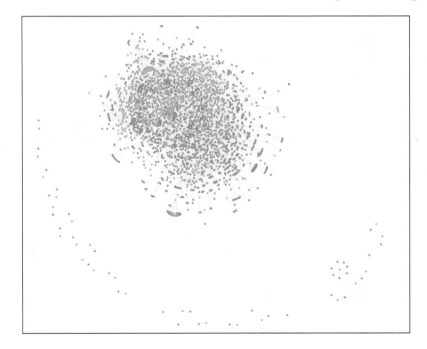

From this picture, we can see that many ingredients share the same compounds, whereas some compounds can only be found in some ingredients. Similar to the social ego network, this network consists of some isolated nodes, and a giant component of connected nodes. This leads to our next topic, which is the measure of the connectedness of graphs.

Finding the connected components

In a network, two nodes are connected if there is a path between them on the graph. A network is called **connected** if all the node pairs are connected. Otherwise, a **disconnected** network has many components, each of which is connected. To find the connected components of a graph is easy in GraphX using the connectedComponents method.

Using the food network as an example, we can verify that it has exactly 27 components:

```
// Connected Components
scala> val cc = foodNetwork.connectedComponents()
cc: org.apache.spark.graphx.Graph[VertexId,Int]

// Number of components
scala> cc.vertices.map(_._2).collect.distinct.size
res: Int = 27
```

Given the type of cc above, we see it returns another graph with the same number of vertices. The vertices belonging to the same component have the same attribute whose value is the smallest vertex ID in that component. In other words, the attribute of each node identifies its component. Let's see these component identifiers:

```
scala> cc.vertices.map(_._2).distinct.collect
res6: Array[org.apache.spark.graphx.VertexId] = Array(892, 0,
1344, 528, 468, 392, 960, 409, 557, 529, 585, 1105, 233, 181, 481,
1146, 970, 622, 1186, 514, 1150, 511, 47, 711, 1211, 1511, 363)
```

Now, suppose we want to list the components and its number of nodes in the descending order. To do this, we can employ Spark's PairedRDD operations which are groupBy and sortBy:

```
scala> cc.vertices.groupBy(_._2).
   map((p => (p._1,p._2.size))).
   sortBy(x => x._2, false).collect()
res: Array[(VertexId, Int)] = Array((0,2580), (528,8), (1344,3),
(392,3), (585,3), (481,3), (892,2), (960,2), (409,2), (557,2),
(529,2), (1105,2), (181,2), (970,2), (622,2), (1186,2), (1150,2),
(511,2), (47,2), (711,2), (1511,2), (363,2), (468,1), (233,1),
(1146,1), (514,1), (1211,1))
```

The giant component has 2580 ingredient and compound nodes, among which the node with the smallest vertex ID is 0. In general, we can define a function that takes the graph of connected components, and returns the smallest vertex ID and number of nodes in the largest component, as follows:

```
def largestComponent(cc: Graph[VertexId, Int]): (VertexId, Int) =
cc.vertices.map(x => (x._2,x._1)).
      groupBy(_._1).
      map(p => (p._1,p._2.size)).
      max()(Ordering.by(_._2))
```

In this function, we grouped the vertices of the components graph by the component ID. Then, we mapped each component to a key-value pair where the key is the component ID, and the value is the number of nodes of the component. Finally, we use the reduction operator called max to return the key-value pair, corresponding to the largest component. In the preceding example, we had to pass to the max method two lists of arguments. The first one is always empty, whereas the second one is an implicit and takes an ordering. To sort pairs on the second element, we had to pass the right ordering to max as Ordering.by(_._2):

In addition to GraphX's graph-specific operations, Spark's RDD and pair RDD operations can be very useful to certain tasks. This function is a canonical example of a chain of data processing in Spark, which is entirely done with Spark's RDD and pair RDD operations. For more details, see the API documentation for Spark and GraphX at http://spark.apache.org/docs/1.1.0/api/scala/index.html#org.apache.spark.package.

Counting triangles and computing clustering coefficients

In the following, we will use the Enron email dataset to illustrate the analysis of a graph connectedness with counting triangle and the clustering coefficients. A triangle is a connected subgraph of three nodes. Counting how many triangles pass through each node helps to quantify the connectedness of graphs. In particular, counting triangle is required to compute the clustering coefficients, which measure the local density of the neighborhood of each node in the network.

Currently, there is a restriction imposed by the triangle counting implementation in Spark on the input graph. Specifically, the edges of the input graph should be in a canonical direction; that is, the `sourceId` parameter must be less than the `destId` parameter. For the email graph, this implies that there should be at most one directed link between the two people. This restriction is not that severe since we can still assume that each directed link in the email graph implies a bidirectional communication between the two people. We can impose this constraint by filtering out the edges for which the ID of the source is larger than that of the destination node. In addition to this restriction, the input graph must also have been partitioned with `partitionBy`. Thus, we can load the email graph as:

```
val emailGraph =
GraphLoader.edgeListFile(sc,"./data/emailEnron.txt").
subgraph(epred = t => t.srcId < t.dstId).
partitionBy(PartitionStrategy.RandomVertexCut)
```

Once the Enron email graph is loaded, we can compute the triangle counts:

```
scala> emailGraph.triangleCount()
res: Graph[Int,Int]
scala> val triangleCounts = emailGraph.triangleCount().vertices
triangleCounts:VertexRDD[Int]
```

Similar to `connectedComponent`, the `triangleCount` algorithm also returns a new graph with the same number of nodes. However, the triangle count becomes the vertex attribute.

How easy was that? Now, let's calculate the local clustering coefficients of the email network. First, we define a function that calculates the local clustering coefficient of a specific node. The clustering coefficient, at a given node, captures the network's local density at that node. The more densely interconnected its neighborhood is, the closer to 1 is its local clustering coefficient. It can be calculated by the following function:

```
def clusterCoeff(tup: (VertexId, (Int,Int))): (VertexId, Double) =
tup match {case (vid, (t, d)) =>
(vid, (2*t.toDouble/(d*(d-1))))
}
```

The argument of `clusterCoeff` is a tuple whose elements consist of the vertex ID of the node at which we compute the local cluster coefficient, and of another tuple, containing the triangle count and degree of the node. Then, it returns the cluster coefficient with the vertex ID as a tuple. Actually, the local cluster coefficient of a given node is an estimate of the probability that each pair of its neighbors is connected. Therefore, the coefficient can be calculated as the ratio between the total links between the node's neighbors, which is also equal to the number of triangles that pass through the node, and the number of all possible pairs of neighboring nodes.

With this, we can compute the cluster coefficients for all the nodes:

```
def clusterCoefficients(graph: Graph[Int,Int]):
RDD[(VertexId, Double)] = {
val gRDD: RDD[(VertexId, (Int, Int))] =
graph.triangleCount().vertices join graph.degrees
gRDD map clusterCoeff
}
```

This last function takes a graph as an input, and returns a pair of RDD collections, whose elements contain the vertex identifiers and the corresponding local coefficients.

The formula for the local clustering coefficient at a given node is well-defined only when its degree, that is the number of its neighbors, is larger than one. If the node has a degree of one or zero, the clusterCoeff function will return a NaN value for the clustering coefficient. Therefore, we must first check if some nodes are isolated in the network when we want to compute an average or global clustering coefficient for a network. Not only must we filter out the leaves and isolated nodes but also, we must adjust the formula of the global clustering coefficient to avoid a biased assessment of the neighborhood clustering.

Let's now use the previous functions to compute the cluster coefficients for the email graph:

```
scala> val coeffs = clusterCoefficients(emailGraph)
scala> coeffs.take(10)
res: Array[(VertexId, Double)] = Array((19021,0.9), (28730,1.0),
(23776,1.0), (31037,0.0), (34207,NaN), (29127,1.0), (9831,NaN),
(5354,0.0380952380952381), (32676,0.46153846153846156), (4926,1.0))
```

We see that for some nodes, the returned clustering coefficient has a NaN value. In fact, this is the case for 25481 out of the 36692 nodes:

```
// Check the NaN values.
scala> coeffs.filter (x => !x._2.isNaN).count
res: Long = 25481
```

To remedy this fact, we need to filter out these nodes when averaging the cluster coefficients:

```
// Calculate the adjusted global cluster coefficient
scala> val nonIsolatedNodes = coeffs.filter(x => !x._2.isNaN)
nonIsolatedNodes: RDD[(VertexId, Double)]
scala> val globalCoeff =
    nonIsolatedNodes.map(_._2).sum / nonIsolatedNodes.count globalCoeff:
Double = 0.7156424032347621
```

The network centrality and PageRank

Previously, we have used the degree distribution and clustering coefficients of a network to understand how connected a network is. In particular, we have learned how to find the largest connected components and the nodes that have the highest degree. Then, we visualized the networks and saw the nodes that have higher chances to play the role of hubs in the network since many nodes are connected to them. In some sense, the degree of a node can be interpreted as a centrality measure that determines how important that node is relative to the rest of the network. In this section, we are going to introduce a different centrality measure as well as the PageRank algorithm, which is useful for ranking nodes in graphs.

There exist many other measures of centrality for graphs. For example, betweenness centrality is useful for information flow problems. Given a node, its betweenness is the number of shortest paths from all vertices to all others that pass through this node. In contrast to PageRank, betweenness centrality assigns a high score to the nodes that are strategically connected on the shortest paths, connecting the pairs of other nodes. Other measures are the connected centrality and Katz centrality. There are no predefined algorithms in GraphX to compute these measures. One of the reasons is the greater complexity required for exactly computing the betweenness centrality. Therefore, approximate algorithms still need to be developed and will be an excellent open source contribution for extending the current GraphX library.

How PageRank works

PageRank is the famous algorithm behind Google's incredibly successful web search engine. In response to each search query, Google wants to display important web pages first. In brief, PageRank assigns a probability score to each page. The higher the score for a node, the more likely a user will land on that page in the long term.

To find the final PageRank scores, the algorithm simulates the behavior of a random surfer by walking her through the web graph. At each step, the surfer can either visit a page that it links to or jump to another random page (this is not necessarily a neighboring page). This is done according to the transition probabilities that are specified by the structure of the graph. For example, a web graph with one thousand nodes will be associated to a 1000 by 1000 transition probability matrix. The element in row i and column j of that matrix has a value of $1/k$ where the j page has k outgoing links, and one of them is to the i page. Otherwise, it is zero. The PageRank algorithm starts at a random node and, at each step, the PageRank scores are updated.

A sketch implementation of this algorithm is shown as follows:

```
var PR = Array.fill(n)(1.0)
val oldPR = Array.fill(n)(1.0)
while( iter <= maxIter || max(abs(PR - oldPr)) > tol) {
  swap(oldPR, PR)
  for(i <- 0 until n) {
    PR[i] = d + (1 - d) * inNbrs[i].map(j => oldPR[j] /
    outDeg[j]).sum
  }
}
```

In the preceding code, alpha is the random reset probability with a default value of `0.15`. Next, `inNbrs[i]` is the set of neighbors, which link to `i`, and `outDeg[j]` is the out-degree of the `j` vertex.

The first term in the update is due to the fact that the surfer can choose to skip the neighbor and instead jump to a random page with a probability as `d`. The second term updates the important score of each page, based on the previous scores of the neighbors that link to the page. This process is repeated until the PageRank scores converge to a fixed value, or until a maximum number of iterations are reached.

In GraphX, there are two implementations of the PageRank algorithm. The first implementation uses the `Pregel` interface and runs PageRank for a fixed number of iterations `numIter`. The second one uses the standalone `Graph` interface and runs PageRank until the change in PageRank score is smaller than a specific error tolerance `tol`.

These PageRank algorithms exploit data-parallelization over vertices. In particular, the `Pregel` implementation relies on local message passing for updating the PageRank scores. Another point to note is that the PageRank scores that are returned are not normalized. Thus, they do not represent a probability distribution. Moreover, pages that have no incoming links will have a PageRank score of alpha. Nonetheless, the top pages can be still be found by sorting the vertices of the returned PageRank graph by their `score` attribute.

Ranking web pages

Here, we will use a new dataset for demonstrating PageRank. The first one is a web graph of pages from the University of Notre Dame. Directed edges represent hyperlinks between them. Using PageRank, we will rank and find the most important pages.

The dataset can be downloaded from `http://snap.stanford.edu/data/web-NotreDame.html`, which was first used by (Albert, Jeong & Barabasi, 1999):

```
// Load web graph
val webGraph = GraphLoader.edgeListFile(sc,"./data/web-NotreDame.txt")

// Run PageRank with an error tolerance of 0.0001
val ranks = webGraph.pageRank(0.001).vertices

// Find the top 10 pages of University of Notre Dame
val topPages = ranks.sortBy(_._2, false).take(10)
```

Scala Build Tool revisited

Previously, we have used the Scala console to interact with Spark. If we want to build a standalone application instead, it becomes unwieldy to manually manage the third-party library dependencies. Remember that first we had to download the JAR files for GraphStream and BreezeViz, as well as those of the libraries that they depend on. Then, we had to put them in the /lib folder and specify this list of JAR files when we submitted the Spark application using the --jars option. This process becomes extremely cumbersome when the application reuses many third-party libraries, which may also depend on several libraries. Fortunately, we can automate this process with SBT. Let's see how to manage the library dependencies, and how to create an uber JAR or assembly JAR with SBT. If you already know how to do this, feel free to skip this section and go ahead to the next chapter.

Organizing build definitions

SBT offers flexibility and power in defining builds and tracking library dependencies. In addition, SBT makes the build process reproducible and interactive. Despite this flexibility, learning all its features can be very discouraging to the unfamiliar user. Instead, we will focus on the essentials.

First, SBT assumes the same directory structure as Maven for the Spark project's source files, which is as follows:

```
src/
  main/
    resources/
<files to include in main jar here>
    scala/
<main Scala sources>
  test/
```

```
    resources
<files to include in test jar here>
    scala/
<test Scala sources>
```

These paths are relative to the project's base directory. On the other hand, build definitions can be put in different files, and can be organized recursively within the project structure. Specifically, there are three places where we can put build definitions:

- A multi-project .sbt build file is recommended in situations where multiple related projects share common settings and dependencies which can be defined in a single build.

- The bare .sbt build files are useful for simple projects. Each .sbt build file defines a list of build and project settings.

- The .scala builds files are combined with the .sbt files to form the complete build definition. Prior to SBT 0.13, this was the old way to share common settings between the multiple projects.

In this book, we will work on simple projects, and the bare .sbt build files will suffice. For details about the mentioned options, refer to the tutorial at http://www.scala-sbt.org/0.13/tutorial/Basic-Def.html.

Managing library dependencies

We can manage library dependencies manually or automatically. In manual mode, we will have to download all libraries in the dependency graph, and then manually copy them in the lib folder. In automatic mode, SBT handles all the work for us by leveraging Apache Ivy mechanisms behind the scenes. With this second method, we need to define three important settings in an SBT build file:

- **Dependencies**: These are the libraries that our application depends on

- **Resolvers**: These are the repositories' locations where SBT will look for the JAR files of these libraries

- **SBT**: These are the plugin settings

> The third set of settings is needed if we want to extend the build definitions using SBT plugins. For example, we will use the sbt-assembly plugin to package a Spark application and the JAR files, it depends on, into a single "uber JAR" file. For this, we need to specify some extra settings such as the uber JAR name as well as the options for creating the uber JAR.

Once we have declared these settings, SBT will take care of the rest for us. Let's look at a concrete example to make sense of all this. We are going to build a Spark application that loads and visualizes food ingredient networks. Earlier in this chapter, we have used the Spark shell and manually managed the dependencies. This time, we will create a standalone application and handle the dependencies automatically.

A preview of the steps

As a preview, here are the steps that we will take to build the Spark application:

1. Create the `plugins.sbt` file inside the `/project` folder. Specify the `sbt-assembly` plugin in that file.

2. Create a `build.sbt` file in the base directory, and declare the project settings.

3. Specify the library dependencies and resolvers.

4. Set up the `sbt-assembly` plugin.

5. Use the SBT commands to assemble the uber JAR.

Step 1 – Enable the sbt-assembly plugin

First, let's enable the `sbt-assembly` plugin. This plugin creates a single deployable uber JAR that contains our built application, and all the libraries that we depend on (except some that we will intentionally exclude from the build). So, let's create the `plugins.sbt` file inside a `new/project` folder. The filename is not important, but it has to be inside the `/project` folder. Then, add this line in the file:

```
addSbtPlugin("com.eed3si9n" % "sbt-assembly" % "0.12.0")
```

Step 2 – Create a build.sbt file

Now, create another `.sbt` file and put it in the base directory. Let's give it a meaningful name, say, `build.sbt` file. As mentioned before, this single file will suffice for our simple project. For more complex ones, it is okay to put the definitions in the multiple `.sbt` files.

As we did in *Chapter 1, Getting Started with Spark and GraphX*, the first things we define in `build.sbt` are the project settings, that is the project name, its version, and the Scala version under which we will build the project. Add the following lines in `build.sbt`:

```
name := "Simple Visualization"

version := "1.0"

scalaVersion := "2.10.4"
```

The `build.sbt` file defines a sequence of build settings. Each element in the sequence is a key-value pair of type `Setting[T]`, where `T` is the expected value type. Each line in `build.sbt` is then a Scala expression, which becomes one element in the sequence called `Seq[Setting[_]]`. For instance, in the expression `name:= "Simple Visualization"`, the left-hand `name` is a key that has a type `SettingKey[String]`. Each key has a method called `:=`, which returns a `Setting[T]`. In our example, the return type of the full expression `name := "Simple Visualization"` is thus `Setting[String]`. In fact, this Scala expression is a syntactic sugar for the method call `name—:=("Simple Visualization")`.

> Do not forget to add empty lines between each setting. Since SBT uses a domain-specific language, the empty lines are mandatory to delineate the build expressions. These blank lines will no longer be needed after the release 0.13.7.

Step 3 – Declare library dependencies and resolvers

To manage the third-party libraries, we will need to attach these libraries to the key called `libraryDependencies` in `build.sbt`. Since an application depends on more than one library, the value type corresponding to `libraryDependencies` is a sequence. Therefore, `libraryDependencies` accepts the append method `+=` to append a dependency, or the concatenate method `++=` to add a list of dependencies. However, it does not accept the operator `:=`.

Our application depends on Spark Core, GraphX, GraphStream, and Breeze libraries. In `build.sbt`, we will attach a list of dependencies, which are as follows:

```
libraryDependencies ++= Seq(
  "org.apache.spark" %% "spark-core" % "1.1.0" % "provided",
  "org.apache.spark" %% "spark-graphx" % "1.1.0" % "provided",
  "org.graphstream" % "gs-core" % "1.2+",
  "org.graphstream" % "gs-ui" % "1.2+",
  "org.scalanlp" % "breeze-viz_2.10" % "0.9",
  "org.scalanlp" % "breeze_2.10" % "0.9"
)
```

Each sequence element in the right-hand side is a Scala expression that returns a `ModuleID` object. Each `ModuleID` object is constructed like this—`groupID % artifactID % revision`. The `groupID`, `artifactID`, and `revision` objects are all `String` objects.

In short, the % method creates the `ModuleID` objects from the passed strings, then we attach those `ModuleID` objects to the setting key `libraryDependencies`.

> Each dependency must correspond with the version of Scala that you are using. For libraries that were built with SBT, such as `spark-core` and `spark-graphx`, we can use the operator `%%` instead of `%` as `groupID %% artifactID % revision`. This will use the right JAR for the dependency, built with the same version of Scala that you are using.
>
> We can also add configuration information to the `ModuleID` like this:
>
>   groupID % artifactID % revision % configuration
>
> For example, in `"org.apache.spark" %% "spark-core" % "1.1.0" % "provided"`, the configuration provided will inform the plugin `sbt-assembly` to exclude JAR files when packaging the uber JAR.

Sometimes, there are pathological cases where two libraries depend on the same library with different versions, and SBT cannot resolve the dependency conflict. For instance, if you try to package and run the application with the `build.sbt` definition, you will get an error like this due to the unresolved dependencies:

```
[error] (*:assembly)  deduplicate: different file contents found
in the following:
~/.ivy2/cache/org.jfree/jfreechart/jars/jfreechart-
1.0.14.jar:org/jfree/chart/ChartPanel.class
~/.ivy2/cache/jfree/jfreechart/jars/jfreechart-
1.0.13.jar:org/jfree/chart/ChartPanel.class
```

This error occurs because both the GraphStream and BreezeViz libraries depend on the Java libraries JFreeChart and JCommon. However, BreezeViz is rarely maintained and is stuck with the `jfreechart-1.0.13` library. To fix this, we have to exclude one of every duplicate JARs. To exclude specific JARs in the dependency graph of a given library, we call one of the methods `exclude` and `excludeAll` on the `ModuleID` object. In our case, we replace the `"org.scalanlp" % "breeze-viz_2.10" % "0.9"` expression by:

```
("org.scalanlp" % "breeze-viz_2.10" % "0.9").
    exclude("jfree","jfreechart").
    exclude("jfree","jcommon")
```

The `exclude` method returns a new `ModuleID` object, but will not include the passed libraries in the final build.

After setting the dependencies, we have to tell SBT where it can download them. This is similarly done by attaching a sequence of repositories to the `resolvers` key as follows:

```
resolvers ++= Seq(
    "Akka Repository" at "http://repo.akka.io/releases/",
    "Sonatype OSS Snapshots" at
"https://oss.sonatype.org/content/repositories/snapshots",
    "Sonatype Releases" at
"http://oss.sonatype.org/content/repositories/releases")
```

Each repository is declared using the form called `name` at location, where the method is invoked on the `String` objects. By default, SBT combines these declared resolvers with the default ones, such as Maven Central or a local Ivy repository.

Step 4 – Set up the sbt-assembly plugin

Next, let's configure the settings of the `sbt-assembly` plugin. Put the following in `build.sbt`:

```
jarName in assembly := "graph-Viz-assembly.jar"
```

This configures the name of the uber JAR or assembly JAR to `graph-Viz-assembly.jar`.

We also need to exclude all the classes from the Scala language distribution. To do this, we tell SBT to exclude all the JARs that either start with `"scala-"`, or are part of the Scala distribution:

```
assemblyOption in assembly := (assemblyOption in
assembly).value.copy(includeScala = false)
```

After this step, `build.sbt` will finally look like this:

```
name := "Simple Visualization"

version := "1.0"

scalaVersion := "2.10.4"

libraryDependencies ++= Seq(
  "org.apache.spark" %% "spark-core" % "1.1.0" % "provided",
  "org.apache.spark" %% "spark-graphx" % "1.1.0" % "provided",
  "org.graphstream" % "gs-core" % "1.2+",
```

```
  "org.graphstream" % "gs-ui" % "1.2+",
  ("org.scalanlp" % "breeze-viz_2.10" % "0.9").
exclude("jfree","jfreechart").exclude("jfree","jcommon"),
  "org.scalanlp" % "breeze_2.10" % "0.9"
)

resolvers ++= Seq(
  "Akka Repository" at "http://repo.akka.io/releases/",
  "Sonatype OSS Snapshots" at "https://oss.sonatype.org/content/
repositories/snapshots",
  "Sonatype Releases" at "http://oss.sonatype.org/content/
repositories/releases")

// Configure jar named used with the assembly plug-in
jarName in assembly := "graph-Viz-assembly.jar"

// Exclude Scala library (JARs that start with scala- and are
included in the binary Scala distribution)
assemblyOption in assembly := (assemblyOption in
assembly).value.copy(includeScala = false)
```

Step 5 – Create the uber JAR

All that needs to be done now is to run the command called `sbt assembly` in the console to build the uber JAR. This must be done with the current directory set to the project base directory:

```
sbt clean assembly
```

This will create the uber JAR within the `target/scala-2.10/` folder. You can look inside the built uber JAR to see all the classes that it contains, which are as follows:

```
jar tf target/scala-2.10/graph-Viz-assembly.jar
```

Finally, we can submit the built application with the `spark-submit` script by passing the assembly JAR this time:

```
../../bin/spark-submit --class
com.github.giocode.graphxbook.SimpleGraphVizApp --master local
target/scala-2.10/graph-Viz-assembly.jar
```

Running tasks with SBT commands

SBT provides different, useful commands for interacting with the build in the SBT console. These are listed as follows:

- `clean`: This removes the files that were previously produced by the build, such as generated sources, compiled classes, and task caches
- `console`: This starts the Scala shell with the project classes on the classpath
- `compile`: This command compiles the sources
- `update`: The execution of this command resolves and retrieves the dependencies, if required
- `package`: This builds and produces a deployable JAR
- `assembly`: This builds a uber JAR using the `sbt-assembly` plugin

Summary

In this chapter, we learned about the different ways to visualize and analyze graphs in Spark. We studied the connectedness of different networks by looking at their degree distribution, finding their connected components, and by calculating their cluster coefficients. In addition, we also learned how to visualize graph data using GraphStream. After this, we showed how the PageRank algorithm can be used to rank node importance in different networks. This chapter also showed us how to use SBT to build a Spark program that relies on third-party libraries.

Throughout this chapter, we have also studied how the basic Spark RDD operations can be used to transform, join, and filter collections of graph vertices and edges. In the next chapter, we will learn about the graph-specific and higher-level operations that are used to transform and manipulate the structure of graphs.

In the next chapter, we will learn about graph-specific operators that help change the properties of graph elements or modify the graph structure.

4
Transforming and Shaping Up Graphs to Your Needs

In this chapter, we will learn to transform graphs using different sets of operators. In particular, we will cover graph-specific operators that either change the properties of graph elements or modify the structure of graphs. In other words, all the operators that we use here are methods that are invoked on a graph and return a new graph. In addition, we will use join methods to combine graph data with other datasets. Using real-world datasets, you will understand when and how to:

- Use property operators to modify vertex or edge properties
- Use structural operators to modify the shape of a graph
- Join additional RDD collections with a property graph

Transforming the vertex and edge attributes

The map operator is a core method for transforming distributed datasets or RDDs in Spark. Similarly, property graphs also have three map operators defined as follows:

```
class Graph[VD, ED] {
  def mapVertices[VD2](mapFun: (VertexId, VD) => VD2): Graph[VD2, ED]
  def mapEdges[ED2](mapFun: Edge[ED] => ED2): Graph[VD, ED2]
  def mapTriplets[ED2](mapFun: EdgeTriplet[VD, ED] => ED2): Graph[VD, ED2]
}
```

Each of these methods is called on a property graph with vertex attribute type VD and edge attribute type ED. Each of them also takes a user-defined mapping function mapFun that performs one of the following:

- For mapVertices, mapFun takes a pair of (VertexId, VD) as input and returns a transformed vertex attribute of type VD2.

- For mapEdges, mapFun takes an Edge object as input and returns a transformed edge attribute of type ED2.

- For mapTriplets, mapFun takes an EdgeTriplet object as input and returns a transformed edge attribute of type ED2.

In each case, the graph structure remains intact, meaning these map operators never change the links between the vertices or their vertex indices. This is one key advantage of these operators compared to the basic RDD map operator. Although the latter can be used to achieve the same result, the former is also more efficient, thanks to the GraphX system optimization. Therefore, these three mapping operators should always be used if you just want to transform a graph's attributes without modifying its structure.

The difference between mapEdges and mapTriplets is that, for the latter, both the edge and source attributes are available in the triplet input of mapFun to create a new edge attribute. In contrast, the mapFun in mapEdges has access to only the edge attribute.

Now, let's see them in action through some simple examples.

mapVertices

Consider a social graph between people, where the vertex attribute has a type Person and the edge attribute has a type Link. First, let's create these Scala types as follows:

```
case class Person(first: String, last: String, age: Int)
case class Link(relationship: String, duration: Float)
```

Suppose we build the graph from VertexRDD called people and an EdgeRDD collection named links:

```
val inputGraph: Graph[Person, Link] = Graph(people, links)
```

If we want, we can transform the attributes of the people to contain only their name using mapVertices:

```
val outputGraph: Graph[String, Link] =
inputGraph.mapVertices((_, person) => person.first + person.last)
```

The new `outputGraph` now has a vertex attribute of type `String` instead of `Person`. The links between the people remain unchanged.

mapEdges

Similarly, suppose we are interested only in the nature of relationships, not their duration. This time, we can use `mapEdges` to change the edge attribute as follows:

```
val outputGraph: Graph[Person, String] =
inputGraph.mapEdges(link => link.relationship)
```

mapTriplets

Finally, suppose we want to keep track of the people's ages from when they first met and add this information into the edge attribute. We can do that by using `mapTriplets`:

```
val outputGraph: Graph[Person, (Int, Int)] =
inputGraph.mapTriplets(t => (t.srcAttr.age - t.attr.duration,
t.dstAttr.age - t.attr.duration))
```

If we want to change both the edge and vertex attributes of a graph, we can simply chain `mapEdges` or `mapTriplets` with `mapVertices` since each of these methods always returns a property graph.

Modifying graph structures

The GraphX library also comes with four useful methods for changing the structure of graphs. Their method signatures are listed as follows:

```
class Graph[VD, ED] {
  def reverse: Graph[VD, ED]

  def subgraph(epred: EdgeTriplet[VD,ED] => Boolean,
               vpred: (VertexId, VD) => Boolean): Graph[VD, ED]

  def mask[VD2, ED2](other: Graph[VD2, ED2]): Graph[VD, ED]

  def groupEdges(merge: (ED, ED) => ED): Graph[VD,ED]
}
```

The reverse operator

As its name suggests, the reverse operator returns a new graph with all the edge directions reversed. It does not modify any vertex or edge properties, or change the number of edges. Moreover, its implementation does not produce any data movement or duplication.

The subgraph operator

Next on the list is subgraph, which is useful for filtering graphs. It takes two predicate functions as arguments that return Boolean values. The first predicate epred takes an EdgeTriplet and returns true when the triplet satisfies the predicate. Similarly, the vpred predicate takes a pair of (VertexId, VD) and returns true when the vertex satisfies the predicate condition.

Using these predicates, subgraph returns the graph containing only the nodes that satisfy the vertex predicate and keeps only the edges satisfying the edge predicate between the remaining nodes. By default, the vertex or edge predicate functions are set to return true when they are not provided. That means that we can pass only an edge predicate, only a vertex predicate, or both. If only a vertex predicate is passed to subgraph and two connected vertices are filtered out, then the edge connecting these nodes will automatically be filtered out as well.

The subgraph operator is very handy for countless situations. For instance, it is often the case, in practice, that the graphs have isolated nodes or edges with missing vertex information. We can eliminate these graph elements using subgraph. Another situation where subgraph is useful is when we want to remove hubs in the graph, for example, nodes that are connected to too many nodes.

As a concrete example, let's use subgraph to answer the following question often encountered in social networks: "Which people in my friends' list of friends are not yet my friends?":

```
// Given a social network
type Name = String
class Person(name: Name, friends: List[Name])
val socialNetwork: Graph[Person, Int] = ...

// that I am part of
val me = Person(myName, myFriends)

// I want know my friends' friends that are not yet my friends
val potentialFriends = socialNetwork.subgraph(vpred =
(_, p: Person) => !(me.friends contains p.name))
```

Note that we did not pass an edge predicate as an argument to subgraph. Thus, Scala uses the default value for epred, which is a function that always returns true. On the other hand, we should pass vpred as a named parameter so that Scala knows which predicate is passed or is missing.

The mask operator

The mask operator also filters a graph on which it is invoked. In contrast to subgraph, mask does not take predicate functions as arguments. Instead, it takes another graph. Then, the expression graph.mask(anotherGraph) constructs a subgraph of graph by returning a graph that contains the vertices and edges that are also found in anotherGraph. This can be used together with the subgraph operator to filter a graph based on the properties in another related graph.

Consider the following situation where we want to find the connected components of a graph but we want to remove vertices with missing attribute information in the resulting graph. We can then run the connectedComponent algorithm we previously saw and use subgraph and mask together to obtain the desired result. This is shown in the following code:

```
// Run Connected Components
val ccGraph = graph.connectedComponents()

// Remove vertices with missing attribute values and the edges
connected to them
val validGraph = graph.subgraph(vpred =
(_, attr) => attr.info != "NA")

// Restrict the resulting components to the valid subgraph
val validCCGraph = ccGraph.mask(validGraph)
```

The groupEdges operator

Spark's property graphs are allowed to pair any of the connected nodes to have multiple edges. The groupEdges operator is another structural operator that merges duplicate edges between each pair of nodes into a single edge. To do that, groupEdges requires one function argument named merge, which takes a pair of edge attributes of type ED and combines them into a single attribute value of the same type. As a result, the graph returned by groupEdges has the same type as the original one. Later in this chapter, we will work on a detailed example in which we will see groupEdges in action.

Joining graph datasets

In addition to the previous mapping and filtering operations, GraphX also provides APIs for joining RDD datasets with graphs. This can be useful when we want to add extra information to the vertex attributes of a graph or when we want to merge the vertex attributes of two related graphs. These tasks can be accomplished using the following join operators.

joinVertices

The following is the method signature for the first operator `joinVertices`:

```
def joinVertices[U](table: RDD[(VertexId, U)])(map: (VertexId, VD,
U) => VD): Graph[VD, ED]
```

It is invoked on a `Graph[VD, ED]` object and requires two inputs, which are passed as curried parameters. First, `joinVertices` joins a graph's vertex attributes with an input vertex RDD table of type `RDD[(VertexId, U)]`. Second, a user-defined `map` function is also passed to `joinVertices`. This `map` function joins the original and passed attributes of each vertex into a new attribute. The return type of this new attribute must be the same as the original one. Moreover, vertices without a matching value in the passed RDD retain their original value.

outerJoinVertices

The second join operator is `outerJoinVertices`, which is a more general method than `joinVertices`. Its method signature is shown as follows:

```
def outerJoinVertices[U, VD2](table: RDD[(VertexId, U)])(map:
(VertexId, VD, Option[U]) => VD2): Graph[VD2, ED]
```

While `outerJoinVertices` also expects a vertex RDD and a user-defined `map` function as parameters, the `map` function is allowed to change the vertex attribute type. Furthermore, all vertices in the original graph are transformed even if they are not present in the passed RDD table.

As a result of this, the `map` function takes an `Option` type parameter `Option[U]` instead of simply `U` as in `joinVertices`.

Example – Hollywood movie graph

An example will help illustrate these differences. For that, let's go to Hollywood and build a small graph of movie actors and actresses:

```
scala> val actors: RDD[(VertexId, String)] = sc.parallelize(List(
    (1L, "George Clooney"),(2L, "Julia Stiles"),
    (3L, "Will Smith"), (4L, "Matt Damon"),
    (5L, "Salma Hayek")))
actors: RDD[(VertexId, String)]
```

Two people in the graph will be connected if they appeared in a movie together. Each edge will contain the movie title. Let's load that information into an edge RDD called `movies`:

```
scala>  val movies: RDD[Edge[String]] = sc.parallelize(List(
    Edge(1L,4L,"Ocean's Eleven"),
    Edge(2L, 4L, "Bourne Ultimatum"),
    Edge(3L, 5L, "Wild Wild West"),
    Edge(1L, 5L, "From Dusk Till Dawn"),
    Edge(3L, 4L, "The Legend of Bagger Vance"))
)
movies: RDD[Edge[String]]
```

Now, we can build the movie graph and see what's inside:

```
scala> val movieGraph = Graph(actors, movies)
movieGraph: Graph[String,String]


scala> movieGraph.triplets.foreach(t => println(
t.srcAttr + " & " + t.dstAttr + " appeared in " + t.attr))


George Clooney & Matt Damon appeared in Ocean's Eleven
Julia Stiles & Matt Damon appeared in Bourne Ultimatum
George Clooney & Salma Hayek appeared in From Dusk Till Dawn
Will Smith & Matt Damon appeared in The Legend of Bagger Vance
Will Smith & Salma Hayek appeared in Wild Wild West
```

For now, our vertices contain only the name of each actor/actress:

```scala
scala> movieGraph.vertices.foreach(println)
(2,Julia Stiles)
(1,George Clooney)
(5,Salma Hayek)
(4,Matt Damon)
(3,Will Smith)
```

Suppose we have access to a dataset of actor biographies. For this example, let's quickly load one such dataset into a vertex RDD:

```scala
scala> case class Biography(birthname: String, hometown: String)
defined class Biography


scala> val bio: RDD[(VertexId, Biography)] = sc.parallelize(List(
   (2, Biography("Julia O'Hara Stiles", "NY City, NY, USA")),
   (3, Biography("Willard Christopher Smith Jr.", "Philadelphia, PA,
USA")),
   (4, Biography("Matthew Paige Damon", "Boston, MA, USA")),
   (5, Biography("Salma Valgarma Hayek-Jimenez", "Coatzacoalcos, Veracruz,
Mexico")),
   (6, Biography("José Antonio Domínguez Banderas", "Málaga, Andalucía,
Spain")),
   (7, Biography("Paul William Walker IV", "Glendale, CA, USA"))
))
bio: RDD[(VertexId, Biography)]
```

We are going to use `joinVertices` to join this information to our movie graph. To do that, let's create the user-defined function that appends the hometown of an actor/actress to their name:

```scala
scala> def appendHometown(id: VertexId, name: String, bio: Biography):
String = name + ":"+ bio.hometown
appendHometown: (id: VertexId, name: String, bio: Biography)String
```

Remember for `joinVertices`, the `mapping` function should return a string because that's the vertex attribute type of the original graph, for example `string`. Now, we can join the biography to the vertex attributes of our Hollywood graph:

```scala
scala> val movieJoinedGraph =
movieGraph.joinVertices(bio)(appendHometown)
```

```
movieJoinedGraph: Graph[String,String]
```

```
scala> movieJoinedGraph.vertices.foreach(println)
```

```
(1,George Clooney)
(5,Salma Hayek:Coatzacoalcos, Veracruz, Mexico)
(2,Julia Stiles:NY City, NY, USA)
(4,Matt Damon:Boston, MA, USA)
(3,Will Smith:Philadelphia, PA, USA)
```

Next, let's use `outerJoinVertices` to see the difference. This time, we will directly pass an anonymous `map` function that joins the name and biography, and return this pair as is:

```
scala> val movieOuterJoinedGraph =
movieGraph.outerJoinVertices(bio)((_,name, bio) => (name,bio))
movieOuterJoinedGraph: Graph[(String, Option[Biography]), String]
```

Notice how `outerJoinVertices` changed the vertex attribute type from a `String` to a `tuple (String, Option[Biography])`. Now, let's print the vertices:

```
scala> movieOuterJoinedGraph.vertices.foreach(println)
(1,(George Clooney,None))
(4,(Matt Damon,Some(Biography(Matthew Paige Damon,Boston, MA, USA))))
(5,(Salma Hayek,Some(Biography(Salma Valgarma Hayek-
Jimenez,Coatzacoalcos, Veracruz, Mexico))))
(2,(Julia Stiles,Some(Biography(Julia O'Hara Stiles,NY City, NY, USA))))
(3,(Will Smith,Some(Biography(Willard Christopher Smith Jr.,Philadelphia,
PA, USA))))
```

As mentioned previously, even if there was not a biography of George Clooney in the `bio` dataset passed to `outerJoinVertices`, its new attribute has been changed to `None`, which is a valid instance of the optional type `Option[Biography]`.

Sometimes, it can be convenient to extract the information outside of the optional value. For this we can use the `getOrElse` method defined on `Option[T]` and provide a default new attribute value for the vertices that are not present in the passed vertex RDD:

```
scala> val movieOuterJoinedGraph = movieGraph.outerJoinVertices(bio)((_,
name, bio) =>
(name,bio.getOrElse(Biography("NA","NA"))))
```

```
movieOuterJoinedGraph: Graph[(String, Biography),String]
```

```
scala> movieOuterJoinedGraph.vertices.foreach(println)
(1,(George Clooney,Biography(NA,NA)))
(2,(Julia Stiles,Biography(Julia O'Hara Stiles,NY City, NY, USA)))
(5,(Salma Hayek,Biography(Salma Valgarma Hayek-Jimenez,Coatzacoalcos,
Veracruz, Mexico)))
(4,(Matt Damon,Biography(Matthew Paige Damon,Boston, MA, USA)))
(3,(Will Smith,Biography(Willard Christopher Smith Jr.,Philadelphia, PA,
USA)))
```

Alternatively, it is possible to create a new return type for the joined vertices. For instance, we can create a type `Actor` to generate a new graph of type `Graph[Actor,String]` as follows:

```
scala> case class Actor(name: String, birthname: String, hometown:
String)
defined class Actor
```

```
scala> val movieOuterJoinedGraph = movieGraph.outerJoinVertices(bio)((_,
name, b) => b match {
     case Some(bio) => Actor(name, bio.birthname, bio.hometown)
     case None => Actor(name, "", "")
  })
movieOuterJoinedGraph: Graph[Actor,String]
```

Listing the vertices of the new graph, we get the expected result:

```
scala> movieOuterJoinedGraph.vertices.foreach(println)
(4,Actor(Matt Damon,Matthew Paige Damon,Boston, MA, USA))
(1,Actor(George Clooney,,))
(5,Actor(Salma Hayek,Salma Valgarma Hayek-Jimenez,Coatzacoalcos,
Veracruz, Mexico))
(2,Actor(Julia Stiles,Julia O'Hara Stiles,NY City, NY, USA))
(3,Actor(Will Smith,Willard Christopher Smith Jr.,Philadelphia, PA, USA))
```

Notice that no new vertices will be created for `Antonio Banderas` or `Paul Walker` despite their presence in the `bio` RDD because they do not belong to the original graph.

When calling outerJoinVertices mentioned previously, we have passed the argument map function without type annotation. This is optional as long as the definition of the map function conforms to the expected input and output types.

Although it's possible for the RDD dataset passed to joinVertices or outerJoinVertices to have more than one value for a vertex, only one value will be used. Therefore, it is recommended that the RDD is made to contain unique vertices.

For both joinVertices and outerJoinVertices, the vertices in the output graphs will be the same. Only their vertex attributes will be different. No new vertex will be created as their role is to join information from the passed RDD into existing vertices.

Data operations on VertexRDD and EdgeRDD

All of the operations we've seen previously are graph operations. They are invoked on a graph and they return a new graph object. In this section, we will introduce operations that transform VertexRDD and EdgeRDD collections. The types of these collections are subtypes of RDD[(VertexID, VD)] and RDD[Edge[ED]] respectively.

Mapping VertexRDD and EdgeRDD

First, mapValues takes a map function as input, which transforms each vertex attribute in the VertexRDD. Then, it returns a new VertexRDD object while preserving the original vertex indices. The method mapValues is overloaded so that the map function can take an input with a type VD or (VertexId, VD). The type of the new vertex attributes can be different to VD:

```
def mapValues[VD2](map: VD => VD2): VertexRDD[VD2]
def mapValues[VD2](map: (VertexId, VD) => VD2): VertexRDD[VD2]
```

For illustration, let's take the biographies of the previous Hollywood stars in a VertexRDD collection:

```
scala> val actorsBio = movieJoinedGraph.vertices
actorsBio: VertexRDD[String]

scala> actorsBio.foreach(println)
(4,Matt Damon:Boston, Massachusetts, USA)
(1,George Clooney)
```

```
(5,Salma Hayek:Coatzacoalcos, Veracruz, Mexico)
(3,Will Smith:Philadelphia, Pennsylvania, USA)
(2,Julia Stiles:New York City, New York, USA)
```

Now, we can use `mapValues` to extract their names into a new `VertexRDD` collection:

```
scala> actorsBio.mapValues(s => s.split(':')(0)).foreach(println)
(2,Julia Stiles)
(1,George Clooney)
(5,Salma Hayek)
(4,Matt Damon)
(3,Will Smith)
```

Using the overloaded `mapValues` method, we can include the vertex IDs in the input of the `map` function and still get a similar result:

```
scala> actorsBio.mapValues((vid,s) => s.split(':')(0)).foreach(println)
(1,George Clooney)
(5,Salma Hayek)
(3,Will Smith)
(4,Matt Damon)
(2,Julia Stiles)
```

There is also one `mapValues` method for transforming `EdgeRDD`s:

```
def mapValues[ED2](f: Edge[ED] => ED2): EdgeRDD[ED2]
```

Similarly, `mapValues` changes only the edge attributes. It does not remove or add edges, nor does it modify the direction of the edges.

Filtering VertexRDDs

Using the `filter` method, we can also filter `VertexRDD` collections. While not changing the vertex indexing, `filter` removes the vertices that do not satisfy a user-defined predicate, which is passed to `filter`. Contrary to `mapValues`, `filter` is not overloaded so the type of the predicate must be `(VertexId, VD) => Boolean`. This is summarized as follows:

```
def filter(pred: (VertexId, VD) => Boolean): VertexRDD[VD]
```

In addition to `filter`, the `diff` operation also filters vertices inside a `VertexRDD` collection. It takes another `VertexRDD` set as input and removes vertices from the original set that are also in the input set:

```
def diff(other: VertexRDD[VD]): VertexRDD[VD]
```

> GraphX does not provide a similar filter operation for `EdgeRDD` collections because filtering edges can be directly and efficiently achieved using the graph operation `subgraph`. See the previous section on *Modifying graph structures*.

Joining VertexRDDs

The following join operators are optimized for `VertexRDD` collections:

```
def innerJoin[U, VD2](other: RDD[(VertexId, U)])(f: (VertexId, VD,
U) => VD2): VertexRDD[VD2]

def leftJoin[U, VD2](other: RDD[(VertexId, VD2)])(f: (VertexId,
VD, Option[U]) => VD2): VertexRDD[VD2]
```

The first operator is `innerJoin`, which takes `VertexRDD` and a user-defined function `f` as inputs. Using this function, it joins the attributes of vertices that are present in both the original and input `VertexRDD` sets. In other words, `innerJoin` returns the intersection set of vertices and merges their attributes according to `f`.

So, given the vertex RDD from `movieGraph`, the result of `innerJoin` with the RDD of biographies will not contain `George Clooney`, `Paul Walker` or `José Antonio Domínguez Banderas`:

```
scala> val actors = movieGraph.vertices

actors: VertexRDD[String]


scala> actors.innerJoin(bio)((vid, name, b) => name + " is from " +
b.hometown).foreach(println)

(4,Matt Damon is from Boston, Massachusetts, USA)

(5,Salma Hayek is from Coatzacoalcos, Veracruz, Mexico)

(2,Julia Stiles is from New York City, New York, USA)

(3,Will Smith is from Philadelphia, Pennsylvania, USA)
```

The second operator `leftJoin` is similar to the operator `outerJoinVertices` defined in `Graph[VD,ED]`. It also takes a user-defined function f of type `(VertexId, VD, Option[U]) => VD2)` in addition to an input `VertexRDD` set. The resulting `VertexRDD` will also contain the same vertices as the original `VertexRDD`. Since the third input of the function f is `Option[U]`, it should handle the case when a vertex in the original `VertexRDD` set is not present in the input RDD. Using the previous example, we would do something like:

```scala
scala> actors.leftJoin(bio)((vid, name, b) => b match {
    case Some(bio) => name + " is from " + bio.hometown
    case None => name + "\'s hometown is unknown"
}).foreach(println)


(4,Matt Damon is from Boston, Massachusetts, USA)
(1,George Clooney's hometown is unknown)
(5,Salma Hayek is from Coatzacoalcos, Veracruz, Mexico)
(2,Julia Stiles is from New York City, New York, USA)
(3,Will Smith is from Philadelphia, Pennsylvania, USA)
```

Joining EdgeRDDs

In GraphX, there exists a join operator `innerJoin` for joining two `EdgeRDD`:

```scala
def innerJoin[ED2, ED3](other: EdgeRDD[ED2])(f: (VertexId,
VertexId, ED, ED2) => ED3): EdgeRDD[ED3]
```

It is similar to the `innerJoin` method for `VertexRDD`, except that now its input function has the type: `f: (VertexId, VertexId, ED, ED2) => ED3`. Moreover, `innerJoin` uses the same partitioning strategy as the original `EdgeRDD`.

Reversing edge directions

Previously, we have seen the `reverse` operation that reverses all the edges of graph. When we want to reverse only a subset of edges in a graph, the following `reverse` method defined as `EdgeRDD` objects becomes useful:

```scala
def reverse: EdgeRDD[ED]
```

For instance, we know that graph properties must be directed in Spark. The only way to model a non-directed graph is to add a reverse link for each edge. This can easily be done using the `reverse` operator as follows. First, we extract the edges of the movie graph into the `EdgeRDD` movie:

```scala
scala> val movies = movieGraph.edges
```

```
movies: EdgeRDD[String,String]

scala> movies.foreach(println)
Edge(1,4,Ocean's Eleven)
Edge(3,5,Wild Wild West)
Edge(2,4,Bourne Ultimatum)
Edge(1,5,From Dusk Till Dawn)
Edge(3,4,The Legend of Bagger Vance)
```

Then, we create a new `EdgeRDD` collection with the links reversed. Then, we obtain the bidirected graph using the union of these two `EdgeRDD` collections:

```
scala> val bidirectedGraph = Graph(actors, movies union
       movies.reverse)
```

We can see that this works by printing the new set of edges:

```
scala> bidirectedGraph.edges.foreach(println)
Edge(1,5,From Dusk Till Dawn)
Edge(3,4,The Legend of Bagger Vance)
Edge(3,5,Wild Wild West)
Edge(1,4,Ocean's Eleven)
Edge(2,4,Bourne Ultimatum)
Edge(4,1,Ocean's Eleven)
Edge(4,2,Bourne Ultimatum)
Edge(5,3,Wild Wild West)
Edge(4,3,The Legend of Bagger Vance)
Edge(5,1,From Dusk Till Dawn)
```

> `EdgeRDD[ED]` is a subtype of `RDD[Edge[ED]]` and it organizes the edges in to blocks partitioned using one of the partitioning strategies defined in `PartitionStrategy`. The edge attributes and adjacency structure are stored separately within each partition so that the structure can be reused when only the edge attributes are changed.
>
> In Spark 1.0 and 1.1, the type signature of `EdgeRDD` has been changed `EdgeRDD[ED, VD]` for optimization purposes. Since Spark 1.2, the signature has switched back to the simpler `EdgeRDD[ED]` type definition while implementing the caching optimization in a different way.

Collecting neighboring information

When doing graph computations, we may want to use neighboring information, such as the attributes of neighboring vertices. The two operators, `collectNeighborIds` and `collectNeighbors` explicitly allow us to do that. `collectNeighborIds` collects into a `VertexRDD` only the vertex IDs of each node's neighbors, whereas `collectNeighbors` also collects their attributes:

```
def collectNeighborIds(edgeDirection: EdgeDirection):
VertexRDD[Array[VertexId]]
def collectNeighbors(edgeDirection: EdgeDirection):
VertexRDD[Array[(VertexId, VD)]]
```

These two methods are invoked on a property graph and are passed with `EdgeDirection` as an input. An `EdgeDirection` attribute can take four possible values:

- `Edge.Direction.In`: When this option is specified, each vertex collects only the attributes of neighbors that have an incoming link to it

- `Edge.Direction.Out`: Each vertex collects only the attributes of neighbors that it links to

- `Edge.Direction.Either`: Each vertex collects the attributes of all its neighbors

- `Edge.Direction.Both`: Each vertex collects the attributes of the neighbors with which it has both an incoming edge and outgoing one

For optimal performance, it is best to avoid using these two operators and rewrite the computation using the more generic and efficient `aggregateMessages` operator presented in the next chapter. The efficiency gain can be substantial especially when implementing an iterative graph-parallel algorithm. But for simple graph transformations that are done only once, it is ok to use `collectNeighors` and `collectNeighborIds`.

Example – from food network to flavor pairing

In *Chapter 2, Building and Exploring Graphs*, we presented the food ingredient dataset and built a bipartite graph that connects each food ingredient to its compounds. In the following, we will build another graph, which consists of only food ingredients. A pair of food ingredients is connected in the new graph only if they share at least one compound. We'll call this new graph the flavor network. We can later use this graph to create new recipes by experimenting with new food pairings.

Let's start with the bipartite food network that we built in *Chapter 2, Building and Exploring Graphs*:

```scala
scala> val nodes = ingredients ++ compounds
scala> val foodNetwork = Graph(nodes, links)
foodNetwork: Graph[Node,Int]
```

To create the new flavor network, we need to know which ingredients share some compounds. This can be done by first collecting the ingredient IDs for each compound node in the `foodNetwork` graph. Concretely, we collect and group ingredient IDs that have that same compound into an RDD collection of tuples (compound id, Array[ingredient id]), as follows:

```scala
scala> val similarIngr: RDD[(VertexId, Array[VertexId])] =
foodNetwork.collectNeighborIds(EdgeDirection.In)
similarIngr: RDD[(VertexId, Array[VertexId])]
```

Next, we create a function `pairIngredients` that takes one such tuple of (compound id, Array[ingredient id]) and creates an edge between every pair of ingredients in the array:

```scala
    def pairIngredients(ingPerComp: (VertexId, Array[VertexId])):
    Seq[Edge[Int]] =
        for {
            x <- ingPerComp._2
            y <- ingPerComp._2
            if x != y
        }   yield Edge(x,y,1)
    pairIngredients:
    (ingPerComp:(VertexId,Array[VertexId]))Seq[Edge[Int]]
```

Once we have that, we can create an `EdgeRDD` collection for every pair of ingredients that share the same compounds from the food network, as follows:

```scala
scala> val flavorPairsRDD: RDD[Edge[Int]] = similarIngr flatMap
pairIngredients
flavorPairsRDD: RDD[Edge[Int]]
```

Finally, we can create the new flavor network:

```scala
scala> val flavorNetwork = Graph(ingredients, flavorPairsRDD).cache
flavorNetwork: Graph[Node,Int]
```

Let's print the first 20 triplets in `flavorNetwork`:

```
scala> flavorNetwork.triplets.take(20).foreach(println)
((3,Ingredient(mackerel,fish/seafood)),(9,Ingredient(peanut_butter,plant
derivative)),1)
((3,Ingredient(mackerel,fish/seafood)),(9,Ingredient(peanut_butter,plant
derivative)),1)
((3,Ingredient(mackerel,fish/seafood)),(9,Ingredient(peanut_butter,plant
derivative)),1)
((3,Ingredient(mackerel,fish/seafood)),(9,Ingredient(peanut_butter,plant
derivative)),1)
((3,Ingredient(mackerel,fish/seafood)),(9,Ingredient(peanut_butter,plant
derivative)),1)
((3,Ingredient(mackerel,fish/seafood)),(9,Ingredient(peanut_butter,plant
derivative)),1)
((3,Ingredient(mackerel,fish/seafood)),(9,Ingredient(peanut_butter,plant
derivative)),1)
((3,Ingredient(mackerel,fish/seafood)),(9,Ingredient(peanut_butter,plant
derivative)),1)
((3,Ingredient(mackerel,fish/seafood)),(9,Ingredient(peanut_butter,plant
derivative)),1)
((3,Ingredient(mackerel,fish/seafood)),(9,Ingredient(peanut_butter,plant
derivative)),1)
((3,Ingredient(mackerel,fish/seafood)),(9,Ingredient(peanut_butter,plant
derivative)),1)
((3,Ingredient(mackerel,fish/seafood)),(9,Ingredient(peanut_butter,plant
derivative)),1)
((3,Ingredient(mackerel,fish/seafood)),(9,Ingredient(peanut_butter,plant
derivative)),1)
((3,Ingredient(mackerel,fish/seafood)),(17,Ingredient(red_
bean,vegetable)),1)
((3,Ingredient(mackerel,fish/seafood)),(17,Ingredient(red_
bean,vegetable)),1)
((3,Ingredient(mackerel,fish/seafood)),(17,Ingredient(red_
bean,vegetable)),1)
((3,Ingredient(mackerel,fish/seafood)),(17,Ingredient(red_
bean,vegetable)),1)
((3,Ingredient(mackerel,fish/seafood)),(17,Ingredient(red_
bean,vegetable)),1)
((3,Ingredient(mackerel,fish/seafood)),(17,Ingredient(red_
bean,vegetable)),1)
((3,Ingredient(mackerel,fish/seafood)),(17,Ingredient(red_
bean,vegetable)),1)
```

It seems mackerel, peanut butter and red beans have something in common. Before we try a new recipe, let's slightly modify the network. Notice that duplicate edges are possible when a pair of ingredients share more than one compound. Suppose we want to group parallel edges between each pair of ingredients into a single edge, which contains the number of shared compounds between the two ingredients. We can do that using the `groupEdges` method:

```
val flavorWeightedNetwork =
flavorNetwork.partitionBy(PartitionStrategy.EdgePartition2D).
groupEdges((x,y) => x+y)
flavorWeightedNetwork: Graph[Node,Int]
```

 `groupEdges` requires the graph to be repartitioned because it assumes that identical edges will be co-located on the same partition. Thus, you must call `partitionBy` prior to grouping the edges.

Now, let's print the 20 pairs of ingredients that share the most compounds:

```
scala> flavorWeightedNetwork.triplets.

sortBy(t => t.attr, false).take(20).

foreach(t => println(t.srcAttr.name + " and " + t.dstAttr.name + " share
" + t.attr + " compounds."))


bantu_beer and beer share 227 compounds.

beer and bantu_beer share 227 compounds.

roasted_beef and grilled_beef share 207 compounds.

grilled_beef and roasted_beef share 207 compounds.

grilled_beef and fried_beef share 200 compounds.

fried_beef and grilled_beef share 200 compounds.

beef and roasted_beef share 199 compounds.

beef and grilled_beef share 199 compounds.

beef and raw_beef share 199 compounds.

beef and fried_beef share 199 compounds.

roasted_beef and beef share 199 compounds.

roasted_beef and raw_beef share 199 compounds.

roasted_beef and fried_beef share 199 compounds.

grilled_beef and beef share 199 compounds.

grilled_beef and raw_beef share 199 compounds.

raw_beef and beef share 199 compounds.
```

```
raw_beef and roasted_beef share 199 compounds.
raw_beef and grilled_beef share 199 compounds.
raw_beef and fried_beef share 199 compounds.
fried_beef and beef share 199 compounds.
```

It is not too surprising that roasted beef and grilled beef have lots of things in common. While the example did not teach us much about culinary arts, it showed that we could mix multiple operators to change a graph into a desired form.

Summary

To summarize, GraphX offers several methods and operators for transforming graph elements and modifying its structure. We can use graph-specific operators, which transform a graph into a new one. In addition, we can use special methods that operate on `VertexRDD` and `EdgeRDD` collections. Moreover, we used join methods to combine graph data with other datasets. You can use all these methods to wrangle new graph datasets and put them in to a shape that suits your specific needs.

In the next chapter, you will learn how to create custom graph operators of your own using generic optimized methods, such as `aggregateMessages` and `mapReduceTriplets`.

5
Creating Custom Graph Aggregation Operators

In the previous chapter, we have seen various operations for transforming the elements of a graph and for modifying its structure. Here, we will learn to use a generic and powerful operator named `aggregateMessages` that is useful for aggregating the neighborhood information of all nodes in the graph. In fact, many graph-processing algorithms rely on iteratively accessing the properties of neighboring nodes and adjacent edges. One such example is the PageRank algorithm.

By applying `aggregateMessages` to the NCAA College Basketball datasets, you will be able to:

- Understand the basic mechanisms and patterns of `aggregateMessages`
- Apply it to create custom graph aggregation operations
- Optimize the performance and efficiency of `aggregateMessages`

NCAA College Basketball datasets

We will again learn by doing in this chapter. This time, we will take the NCAA College Basketball as an illustrative example. Specifically, we use two CSV datasets. The first one `teams.csv` contains the list of all college teams that played in the NCAA Division I competition. Each team is associated with a four-digit ID number. The second dataset `stats.csv` contains the score and statistics of every game during the 2014-2015 regular season. Using the techniques learned in *Chapter 2, Building and Exploring Graphs*, let's parse and load these datasets and load them into RDDs:

1. We create a class `GameStats` that records the statistics of one team during a specific basketball game:

```
case class GameStats(
```

```
    val score: Int,
    val fieldGoalMade:    Int,
    val fieldGoalAttempt: Int,
    val threePointerMade: Int,
    val threePointerAttempt: Int,
    val threeThrowsMade: Int,
    val threeThrowsAttempt: Int,
    val offensiveRebound: Int,
    val defensiveRebound: Int,
    val assist: Int,
    val turnOver: Int,
    val steal: Int,
    val block: Int,
    val personalFoul: Int
)
```

2. We also add the following methods to `GameStats` in order to know how efficient a team's offense was during a game:

```
// Field Goal percentage
def fgPercent: Double = 100.0 * fieldGoalMade /
fieldGoalAttempt

// Three Point percentage
def tpPercent: Double = 100.0 * threePointerMade /
threePointerAttempt

// Free throws percentage
def ftPercent: Double = 100.0 * threeThrowsMade /
threeThrowsAttempt
override def toString: String = "Score: " + score
```

3. We now create a couple of classes for the games' result:

```
abstract class GameResult(
    val season:    Int,
    val day:       Int,
    val loc:       String
)

case class FullResult(
    override val season:    Int,
    override val day:       Int,
    override val loc:       String,
    val winnerStats:       GameStats,
    val loserStats:        GameStats
) extends GameResult(season, day, loc)
```

FullResult has the year and day of the season, the location where the game was played, and the game statistics of both the winning and losing teams.

4. We will then create a statistics graph of the regular seasons. In this graph, the nodes are the teams, whereas each edge corresponds to a specific game. To create the graph, let's parse the CSV file teams.csv into the RDD teams:

```scala
val teams: RDD[(VertexId, String)] =
    sc.textFile("./data/teams.csv").
    filter(! _.startsWith("#")).
    map {line =>
        val row = line split ','
        (row(0).toInt, row(1))
    }
```

5. We can check the first few teams in this new RDD:

```scala
scala> teams.take(3).foreach{println}
(1101,Abilene Chr)
(1102,Air Force)
(1103,Akron)
```

6. We do the same thing to obtain an RDD of the game results, which will have a type RDD[Edge[FullResult]]. We just parse stats.csv and record the fields that we need — the ID of the winning team, the ID of the losing team, and the game statistics of both teams:

```scala
val detailedStats: RDD[Edge[FullResult]] =
  sc.textFile("./data/stats.csv").
  filter(! _.startsWith("#")).
  map {line =>
      val row = line split ','
      Edge(row(2).toInt, row(4).toInt,
          FullResult(
              row(0).toInt, row(1).toInt,
              row(6),
              GameStats(
                                score = row(3).toInt,
                        fieldGoalMade = row(8).toInt,
                     fieldGoalAttempt = row(9).toInt,
                     threePointerMade = row(10).toInt,
                  threePointerAttempt = row(11).toInt,
                      threeThrowsMade = row(12).toInt,
                   threeThrowsAttempt = row(13).toInt,
                      offensiveRebound = row(14).toInt,
                      defensiveRebound = row(15).toInt,
                               assist = row(16).toInt,
                             turnOver = row(17).toInt,
```

```
                        steal = row(18).toInt,
                        block = row(19).toInt,
                   personalFoul = row(20).toInt
              ),
              GameStats(
                        score = row(5).toInt,
                fieldGoalMade = row(21).toInt,
             fieldGoalAttempt = row(22).toInt,
             threePointerMade = row(23).toInt,
          threePointerAttempt = row(24).toInt,
              threeThrowsMade = row(25).toInt,
           threeThrowsAttempt = row(26).toInt,
             offensiveRebound = row(27).toInt,
             defensiveRebound = row(28).toInt,
                        assist = row(20).toInt,
                     turnOver = row(30).toInt,
                        steal = row(31).toInt,
                        block = row(32).toInt,
                   personalFoul = row(33).toInt
              )
           )
        )
     }
```

Let's check what we have got:

```
scala> detailedStats.take(3).foreach(println)
Edge(1165,1384,FullResult(2006,8,N,Score: 75-54))
Edge(1393,1126,FullResult(2006,8,H,Score: 68-37))
Edge(1107,1324,FullResult(2006,9,N,Score: 90-73))
```

7. We then create our graph of stats:

```
scala> val scoreGraph = Graph(teams, detailedStats)
```

For curiosity, let's see which team has won against the 2015 NCAA champions Duke in the regular season. To do that, we filter the graph triplets whose destination attribute is Duke. This is because when we created our stats graph, each edge is directed from the winner node to the loser node. So, Duke has lost only four games in the regular season:

```
scala> scoreGraph.triplets.filter(_.dstAttr == "Duke").foreach(println)
((1274,Miami FL),(1181,Duke),FullResult(2015,71,A,Score: 90-74))
((1301,NC State),(1181,Duke),FullResult(2015,69,H,Score: 87-75))
((1323,Notre Dame),(1181,Duke),FullResult(2015,86,H,Score: 77-73))
((1323,Notre Dame),(1181,Duke),FullResult(2015,130,N,Score: 74-64))
```

The aggregateMessages operator

Once we have our graph ready, let's start our mission, which is aggregating the stats data in `scoreGraph`. In GraphX, `aggregateMessages` is the operator for that kind of job.

For example, let's find out the average field goals made per game by the winning teams. In other words, the games that the teams lost will not be counted. To get the average for each team, we first need to have the number of games won by the team and the total field goals that the team made in those games:

```
// Aggregate the total field goals made by winning teams
type FGMsg = (Int, Int)
val winningFieldGoalMade: VertexRDD[FGMsg] = scoreGraph
aggregateMessages(
    // sendMsg
    triplet => triplet.sendToSrc(1,
    triplet.attr.winnerStats.fieldGoalMade)
    // mergeMsg
    ,(x, y) => (x._1 + y._1, x._2+ y._2)
)
// Aggregate the total field goals made by winning teams
type Msg = (Int, Int)
type Context = EdgeContext[String, FullResult, Msg]
val winningFieldGoalMade: VertexRDD[Msg] = scoreGraph
aggregateMessages(
    // sendMsg
    (ec: Context) => ec.sendToSrc(1,
     ec.attr.winnerStats.fieldGoalMade),

    // mergeMsg
    (x: Msg, y: Msg) => (x._1 + y._1, x._2+ y._2)
)
```

EdgeContext

There is a lot going on in the previous call to `aggregateMessages`. So, let's see it working in slow motion. When we called `aggregateMessages` on the `scoreGraph` method, we had to pass two functions as arguments.

The first function has a signature `EdgeContext[VD, ED, Msg] => Unit`. It takes an `EdgeContext` parameter as input. It does not return anything but it can produce side effects, such as sending a message to a node.

Ok, but what is that `EdgeContext` type? Similar to `EdgeTriplet`, `EdgeContext` represents an edge along with its neighboring nodes. It can access both the edge attribute, and the source and destination nodes' attributes. In addition, `EdgeContext` has two methods to send messages along the edge to its source node or to its destination node. These methods are `sendToSrc` and `sendToDst` respectively. Then, the type of message that we want each triplet in the graph to send is defined by `Msg`. Similar to `VD` and `ED`, we can define the concrete type that `Msg` takes.

In our example, we need to aggregate the number of games played and the number of field goals made. Therefore, we define `Msg` as a pair of `Int`. Furthermore, each edge context sends a message to only its source node, that is the winning team, because we are interested in the total field goals made by the teams for only the games that they won. The actual message sent to each winner node is a pair of integers (`1, ec.attr.winnerStats.fieldGoalMade`). The first integer serves as a counter for the games won by the source node, whereas the second one corresponds to the number of field goals made by the winner. This latter integer is then extracted from the edge attribute.

In addition to `sendMsg`, the second function that we need to pass to `aggregateMessages` is a `mergeMsg` function with the signature `(Msg, Msg) =>` `Msg`. As its name implies, `mergeMsg` is used to merge two messages received at each node into a new one. Its output type must be the same, for example `Msg`. Using these two functions, `aggregateMessages` returns the aggregated messages inside `VertexRDD[Msg]`.

Returning to our example, we set out to compute the average field goals per winning game for all teams. To get this final result, we simply apply `mapValues` to the output of `aggregateMessages`, as follows:

```
// Average field goals made per Game by winning teams
val avgWinningFieldGoalMade: VertexRDD[Double] =
    winningFieldGoalMade mapValues (
        (id: VertexId, x: Msg) => x match {
            case (count: Int, total: Int) => total.toDouble/count
})
```

Let's check the output:

```
scala> avgWinningFieldGoalMade.take(5).foreach(println)
(1260,24.71641791044776)
(1410,23.56578947368421)
(1426,26.239436619718308)
(1166,26.137614678899084)
(1434,25.34285714285714)
```

The definitions of `aggregateMessages` and `EdgeContext`, as we explained previously, are shown as follows:

```
class Graph[VD, ED] {
  def aggregateMessages[Msg: ClassTag] (
      sendMsg: EdgeContext[VD, ED, Msg] => Unit,
      mergeMsg: (Msg, Msg) => Msg,
      tripletFields: TripletFields = TripletFields.All)
    : VertexRDD[Msg]
}



abstract class EdgeContext[VD, ED, A] {

    // Attribute associated with the edge:
    abstract def attr: ED

    // Vertex attribute of the edge's source vertex.
    abstract def srcAttr: VD

    // Vertex attribute of the edge's destination vertex.
    abstract def dstAttr: VD

    // Vertex id of the edge's source vertex.
    abstract def srcId: VertexId

    // Vertex id of the edge's destination vertex.
    abstract def dstId: VertexId

    // Sends a message to the destination vertex.
    abstract def sendToDst(msg: A): Unit

    // Sends a message to the source vertex.
    abstract def sendToSrc(msg: A): Unit
}
```

Abstracting out the aggregation

That was kinda cool! We can do the same to average the points per game scored by winning teams:

```
// Aggregate the points scored by winning teams
val winnerTotalPoints: VertexRDD[(Int, Int)] =
scoreGraph.aggregateMessages(
    // sendMsg
```

```
    triplet => triplet.sendToSrc(1,
    triplet.attr.winnerStats.score),
    // mergeMsg
    (x, y) => (x._1 + y._1, x._2+ y._2)
)


// Average field goals made per Game by winning teams
var winnersPPG: VertexRDD[Double] =
        winnerTotalPoints mapValues (
            (id: VertexId, x: (Int, Int)) => x match {
                case (count: Int, total: Int) =>
                total.toDouble/count
            })
```

Let's check the output:

```scala
scala> winnersPPG.take(5).foreach(println)
(1260,71.19402985074628)
(1410,71.11842105263158)
(1426,76.30281690140845)
(1166,76.89449541284404)
(1434,74.28571428571429)
```

Now, the coach wants us to list the top five teams with the highest average three-pointer made per winning game. By the way, he also wants to know which teams are the most efficient in three-pointers.

Keeping things DRY

We can copy and modify the previous code but that would be repetitive. Instead, let's abstract out the average aggregation operator so that it can work on any statistics that the coach needs. Luckily, Scala's higher-order functions are there to help in this task.

For each statistic that our coach wants, let's define a function that takes a team's GameStats as input and returns the statistic that we are interested in. For now, we will need the number of three-pointers made and the average three-pointer percentage:

```
// Getting individual stats
def threePointMade(stats: GameStats) =
stats.threePointerMade
def threePointPercent(stats: GameStats) = stats.tpPercent
```

Then, we create a generic function that takes as inputs a stats graph and one of the functions defined previously, which has a signature `GameStats => Double`:

```
// Generic function for stats averaging
def averageWinnerStat(graph: Graph[String, FullResult])(getStat:
GameStats => Double): VertexRDD[Double] = {
    type Msg = (Int, Double)
    val winningScore: VertexRDD[Msg] =
    graph.aggregateMessages[Msg] (
        // sendMsg
        triplet => triplet.sendToSrc(1,
        getStat(triplet.attr.winnerStats)),
        // mergeMsg
        (x, y) => (x._1 + y._1, x._2+ y._2)
    )
    winningScore mapValues (
        (id: VertexId, x: Msg) => x match {
            case (count: Int, total: Double) => total/count
        })
}
```

Then, we can use the average stats by passing the functions `threePointMade` and `threePointPercent` to `averageWinnerStat`:

```
val winnersThreePointMade =
averageWinnerStat(scoreGraph)(threePointMade)
val winnersThreePointPercent =
averageWinnerStat(scoreGraph)(threePointPercent)
```

With little effort, we can tell the coach which five winning teams scored the highest number of threes per game:

```
scala> winnersThreePointMade.sortBy(_._2,false).take(5).foreach(println)
(1440,11.274336283185841)
(1125,9.521929824561404)
(1407,9.008849557522124)
(1172,8.967441860465117)
(1248,8.915384615384616)
```

While we are at it, let's find out the five most efficient teams in three-pointers:

```
scala> winnersThreePointPercent.sortBy(_._2,false).take(5).
foreach(println)
(1101,46.90555728464225)
(1147,44.224282479431224)
```

```
(1294,43.754532434101534)
(1339,43.52308905887638)
(1176,43.080814169045105)
```

Interestingly, the teams that made the most three-pointers per winning game are not always the ones who are the most efficient at it. But, they still won those games, which is more important.

Coach wants more numbers

The coach seems unsatisfied with that argument and wants us to get the same statistics but wants us to average them over all the games that each team has played.

Thus, we have to aggregate the information from all the nodes of our graph, and not only at the destination nodes. To make our previous abstraction more flexible, let's create the following types:

```
trait Teams
case class Winners extends Teams
case class Losers extends Teams
case class AllTeams extends Teams
```

We modify the previous higher-order function to have an extra argument `Teams`, which will help us specify at which nodes we want to collect and aggregate the required game stats. The new function becomes:

```
def averageStat(graph: Graph[String, FullResult])(getStat:
GameStats => Double, tms: Teams): VertexRDD[Double] = {
    type Msg = (Int, Double)
    val aggrStats: VertexRDD[Msg] = graph.aggregateMessages[Msg](
        // sendMsg
        tms match {
            case _ : Winners => t => t.sendToSrc((1,
            getStat(t.attr.winnerStats)))
            case _ : Losers  => t => t.sendToDst((1,
            getStat(t.attr.loserStats)))
            case _          => t => {
                t.sendToSrc((1, getStat(t.attr.winnerStats)))
                t.sendToDst((1, getStat(t.attr.loserStats)))
            }
        }
        ,
        // mergeMsg
        (x, y) => (x._1 + y._1, x._2+ y._2)
    )

    aggrStats mapValues (
```

```
            (id: VertexId, x: Msg) => x match {
                case (count: Int, total: Double) => total/count
                })
    }
```

Compared to `averageWinnerStat`, `aggregateStat` allows us to choose whether we want to aggregate the stats for winners only, for losers only, or for all teams. Since the coach wants the overall stats averaged over all games played, we aggregate the stats by passing the `AllTeams()` flag in `aggregateStat`. In this case, we simply define the `sendMsg` argument in `aggregateMessages` so that the required stats are sent to both the source (the winner) and to the destination (the loser) using the `EdgeContext` class's `sendToSrc` and `sendToDst` functions respectively. This mechanism is pretty straightforward. We just need to make sure we send the right information to the right node. In this case, we send `winnerStats` to the winner and `loserStats` to the loser.

Ok, you've got the idea now. So, let's apply it to please our coach. Here are the teams with the overall highest three-pointers per page:

```
// Average Three Point Made Per Game for All Teams
val allThreePointMade = averageStat(scoreGraph)(threePointMade,
AllTeams())
```

Let's see the output:

```
scala> allThreePointMade.sortBy(_._2, false).take(5).foreach(println)
(1440,10.180811808118081)
(1125,9.098412698412698)
(1172,8.575657894736842)
(1184,8.428571428571429)
(1407,8.411149825783973)
```

Here are the five most efficient teams overall in three-pointers per game:

```
// Average Three Point Percent for All Teams
val allThreePointPercent =
averageStat(scoreGraph)(threePointPercent, AllTeams())
```

The output is:

```
scala> allThreePointPercent.sortBy(_._2,false).take(5).foreach(println)
(1429,38.8351815824302)
(1323,38.522819895594)
(1181,38.43052051444854)
(1294,38.41227053353959)
(1101,38.097896464168954)
```

Actually, there is only a 2 percent difference between the most efficient team and the one in the fiftieth position. Most NCAA teams are therefore pretty efficient behind the line. I bet the coach knew that already!

Calculating average points per game

We can also reuse the `averageStat` function to get the average points per game for the winners. In particular, let's take a look at the two teams that won games with the highest and lowest scores:

```
// Winning teams
val winnerAvgPPG = averageStat(scoreGraph)(score, Winners())
```

Let's check the output:

```
scala> winnerAvgPPG.max()(Ordering.by(_._2))

res36: (org.apache.spark.graphx.VertexId, Double) =
(1322,90.73333333333333)
```

```
scala> winnerAvgPPG.min()(Ordering.by(_._2))

res39: (org.apache.spark.graphx.VertexId, Double) = (1197,60.5)
```

Apparently, the most defensive team can win games by scoring only 60 points, whereas the most offensive team can score an average of 90 points.

Next, let's average the points per game for all games played and look at the two teams with the best and worst offense during the 2015 season:

```
// Average Points Per Game of All Teams
val allAvgPPG = averageStat(scoreGraph)(score, AllTeams())
```

The output is:

```
scala> allAvgPPG.max()(Ordering.by(_._2))

res42: (org.apache.spark.graphx.VertexId, Double) =
(1322,83.81481481481481)
```

```
scala> allAvgPPG.min()(Ordering.by(_._2))

res43: (org.apache.spark.graphx.VertexId, Double) =
(1212,51.111111111111114)
```

To no surprise, the best offensive team is the same as the one who scored most in winning games. To win a game, 50 points is not enough of an average for a team.

Defense stats – D matters as in direction

Previously, we obtained some statistics such as field goals or the three-point percentages that a team achieves. What if instead we want to aggregate the average points or rebounds that each team concedes to their opponents? To compute that, we define a new higher-order function averageConcededStat. Compared to averageStat, this function needs to send loserStats to the winning team and winnerStats to the losing team. To make things more interesting, we are going to make the team name part of the message Msg:

```
def averageConcededStat(graph: Graph[String, FullResult])(getStat:
GameStats => Double, rxs: Teams): VertexRDD[(String, Double)] = {
    type Msg = (Int, Double, String)
    val aggrStats: VertexRDD[Msg] = graph.aggregateMessages[Msg](
        // sendMsg
        rxs match {
            case _ : Winners => t => t.sendToSrc((1,
            getStat(t.attr.loserStats), t.srcAttr))
            case _ : Losers  => t => t.sendToDst((1,
            getStat(t.attr.winnerStats), t.dstAttr))
            case _           => t => {
                t.sendToSrc((1,
                getStat(t.attr.loserStats),t.srcAttr))
                t.sendToDst((1,
                getStat(t.attr.winnerStats),t.dstAttr))
            }
        }
        ,
        // mergeMsg
        (x, y) => (x._1 + y._1, x._2+ y._2, x._3)
    )

    aggrStats mapValues (
        (id: VertexId, x: Msg) => x match {
            case (count: Int, total: Double, name: String) =>
            (name, total/count)
        })
}
```

With that, we can calculate the average points conceded by the winning and losing teams as follows:

```
val winnersAvgConcededPoints =
averageConcededStat(scoreGraph)(score, Winners())
val losersAvgConcededPoints =
averageConcededStat(scoreGraph)(score, Losers())
```

Let's check the output:

```scala
scala> losersAvgConcededPoints.min()(Ordering.by(_._2))
res: (VertexId, (String, Double)) = (1101,(Abilene
Chr,74.04761904761905))

scala> winnersAvgConcededPoints.min()(Ordering.by(_._2))
res: (org.apache.spark.graphx.VertexId, (String, Double)) =
(1101,(Abilene Chr,74.04761904761905))

scala> losersAvgConcededPoints.max()(Ordering.by(_._2))
res: (VertexId, (String, Double)) = (1464,(Youngstown
St,78.85714285714286))

scala> winnersAvgConcededPoints.max()(Ordering.by(_._2))
res: (VertexId, (String, Double)) = (1464,(Youngstown St,71.125))
```

The previous code tells us that Abilene Christian University is the most defensive team. They concede the least points whether they win a game or not. On the other hand, Youngstown has the worst defense.

Joining average stats into a graph

The previous example shows us how flexible the aggregateMessages operator is. We can define the type Msg of the messages to be aggregated to fit our needs. Moreover, we can select which nodes receive the messages. Finally, we can also define how we want to merge the messages.

As a final example, let's aggregate many statistics about each team and join this information into the nodes of the graph:

1. To start, we create its own class for the team stats:

    ```scala
    // Average Stats of All Teams
    case class TeamStat(
            wins: Int   = 0       // Number of wins
          ,losses: Int  = 0       // Number of losses
            ,ppg: Int   = 0       // Points per game
            ,pcg: Int   = 0       // Points conceded per game
            ,fgp: Double = 0    // Field goal percentage
            ,tpp: Double = 0    // Three point percentage
            ,ftp: Double = 0    // Free Throw percentage
          ){
          override def toString = wins + "-" + losses
    }
    ```

2. We collect the average stats for all teams using `aggregateMessages`. For that, we define the type of the message to be an 8-element tuple that holds the counter for games played, won, lost, and other statistics that will be stored in `TeamStat`, as listed previously:

```
type Msg = (Int, Int, Int, Int, Int, Double, Double,
Double)


val aggrStats: VertexRDD[Msg] =
scoreGraph.aggregateMessages(
  // sendMsg
  t => {
        t.sendToSrc((   1,
                        1, 0,
                        t.attr.winnerStats.score,
                        t.attr.loserStats.score,
                        t.attr.winnerStats.fgPercent,
                        t.attr.winnerStats.tpPercent,
                        t.attr.winnerStats.ftPercent
                    ))
        t.sendToDst((   1,
                        0, 1,
                        t.attr.loserStats.score,
                        t.attr.winnerStats.score,
                        t.attr.loserStats.fgPercent,
                        t.attr.loserStats.tpPercent,
                        t.attr.loserStats.ftPercent
                    ))
  }
  ,
  // mergeMsg
  (x, y) => ( x._1 + y._1, x._2 + y._2,
              x._3 + y._3, x._4 + y._4,
              x._5 + y._5, x._6 + y._6,
              x._7 + y._7, x._8 + y._8
            )
)
```

3. Given the aggregate message `aggrStats`, we map them into a collection of `TeamStats`:

```
val teamStats: VertexRDD[TeamStat] = aggrStats mapValues {
  (id: VertexId, m: Msg) => m match {
      case ( count: Int,
             wins: Int,
             losses: Int,
```

```
                totPts: Int,
                totConcPts: Int,
                totFG: Double,
                totTP: Double,
                totFT: Double)   => TeamStat( wins, losses,
                                              totPts/count,
                                              totConcPts/count,
                                              totFG/count,
                                              totTP/count,
                                              totFT/count)

    }
}
```

4. Let's join `teamStats` into the graph. For that, we first create a class `Team` as a new type for the vertex attribute. `Team` will have the name and the `TeamStat` option:

```
case class Team(name: String, stats: Option[TeamStat]) {
    override def toString = name + ": " + stats
}
```

5. We use the `joinVertices` operator, which we have seen in the previous chapter:

```
// Joining the average stats to vertex attributes
def addTeamStat(id: VertexId, t: Team, stats: TeamStat) =
Team(t.name, Some(stats))

val statsGraph: Graph[Team, FullResult] =
    scoreGraph.mapVertices((_, name) => Team(name, None)).
            joinVertices(teamStats)(addTeamStat)
```

6. We can see that the join has worked well by printing the first three vertices in the new graph `statsGraph`:

```
scala> statsGraph.vertices.take(3).foreach(println)

(1260,Loyola-Chicago: Some(17-13))

(1410,TX Pan American: Some(7-21))

(1426,UT Arlington: Some(15-15))
```

7. To conclude this task, let's find out the top 10 teams in the regular seasons. To do so, we define an `Ordering` option for `Option[TeamStat]` as follows:

```
import scala.math.Ordering
object winsOrdering extends Ordering[Option[TeamStat]] {
    def compare(x: Option[TeamStat], y: Option[TeamStat]) =
    (x, y) match {
        case (None, None)       => 0
```

```
        case (Some(a), None)    => 1
        case (None, Some(b))    => -1
        case (Some(a), Some(b)) => if (a.wins == b.wins)
        a.losses compare b.losses
        else a.wins compare b.wins
    }
}
```

8. Finally:

```
import scala.reflect.classTag
import scala.reflect.ClassTag
scala> statsGraph.vertices.sortBy(v =>
v._2.stats,false)(winsOrdering, classTag[Option[TeamStat]]).
    |
take(10).foreach(println)
(1246,Kentucky: Some(34-0))
(1437,Villanova: Some(32-2))
(1112,Arizona: Some(31-3))
(1458,Wisconsin: Some(31-3))
(1211,Gonzaga: Some(31-2))
(1320,Northern Iowa: Some(30-3))
(1323,Notre Dame: Some(29-5))
(1181,Duke: Some(29-4))
(1438,Virginia: Some(29-3))
(1268,Maryland: Some(27-6))
```

 Note that the ClassTag parameter is required in sortBy to make use of Scala's reflection. That is why we had the previous imports.

Performance optimization

In addition to the sendMsg and mergeMsg methods, aggregateMessages can also take an optional argument TripletFields, which indicates what data is accessed in EdgeContext. The main reason for explicitly specifying such information is to help optimize the performance of the aggregateMessages operation.

In fact, TripletFields represents a subset of the fields of _EdgeTriplet_ and it enables GraphX to populate only those fields that are necessary.

The default value is `TripletFields.All`, which means that the `sendMsg` function may access any of the fields in the `EdgeContext` class. Otherwise, the `TripletFields` argument is used to tell GraphX that only part of `EdgeContext` will be required so that an efficient join strategy can be used. All possible options for the `TripletFields` are listed as follows:

- `TripletFields.All`: This option exposes all the fields (source, edge, and destination)

- `TripletFields.Dst`: This one exposes the destination and edge fields but not the source field

- `TripletFields.EdgeOnly`: This option exposes only the edge field but not the source or destination field

- `TripletFields.None`: With this option none of the triplet fields are exposed

- `TripletFields.Src`: This one exposes the source and edge fields but not the destination field

Using our previous example, if we are interested in computing the total number of wins and losses for each team, we will not need to access any fields of the `EdgeContext` class. In this case, we should use `TripletFields.None` to indicate so:

```
// Number of wins of the teams
val numWins: VertexRDD[Int] = scoreGraph.aggregateMessages(
    triplet => {
        triplet.sendToSrc(1)       // No attribute is passed but an
                                   integer
    },
    (x, y) => x + y,
    TripletFields.None
)

// Number of losses of the teams
val numLosses: VertexRDD[Int] = scoreGraph.aggregateMessages(
    triplet => {
        triplet.sendToDst(1)       // No attribute is passed but an
                                   integer
    },
    (x, y) => x + y,
    TripletFields.None
)
```

To see that this works, let's print the top five and bottom five teams:

```
scala> numWins.sortBy(_._2,false).take(5).foreach(println)
(1246,34)
```

```
(1437,32)
(1112,31)
(1458,31)
(1211,31)

scala> numLosses.sortBy(_._2, false).take(5).foreach(println)
(1363,28)
(1146,27)
(1212,27)
(1197,27)
(1263,27)
```

Should you want the name of the top five teams, you need to access the srcAttr attribute. In this case, we need to set tripletFields to TripletFields.Src.

Kentucky as the undefeated team in the regular season:

```
val numWinsOfTeams: VertexRDD[(String, Int)] = scoreGraph.
aggregateMessages(
    t => {
        t.sendToSrc(t.srcAttr, 1)          // Pass source attribute
only
    },
    (x, y) => (x._1, x._2 + y._2),
    TripletFields.Src
)
```

Et voila!:

```
scala> numWinsOfTeams.sortBy(_._2._2, false).take(5).foreach(println)
(1246,(Kentucky,34))
(1437,(Villanova,32))
(1112,(Arizona,31))
(1458,(Wisconsin,31))
(1211,(Gonzaga,31))

scala> numWinsOfTeams.sortBy(_._2._2).take(5).foreach(println)
(1146,(Cent Arkansas,2))
(1197,(Florida A&M,2))
(1398,(Tennessee St,3))
(1263,(Maine,3))
(1420,(UMBC,4))
```

Kentucky has not lost any of its 34 games during the regular season. Too bad that they could not make it into the championship final.

The MapReduceTriplets operator

Prior to Spark 1.2, there was no `aggregateMessages` method in Graph. Instead, the now deprecated `mapReduceTriplets` was the primary aggregation operator. The API for `mapReduceTriplets` is:

```
class Graph[VD, ED] {
  def mapReduceTriplets[Msg] (
      map: EdgeTriplet[VD, ED] => Iterator[(VertexId, Msg)],
      reduce: (Msg, Msg) => Msg)
    : VertexRDD[Msg]
}
```

Compared to `mapReduceTriplets`, the new operator `aggregateMessages` is more expressive as it employs the message passing mechanism instead of returning an iterator of messages as `mapReduceTriplets` does. In addition, `aggregateMessages` explicitly requires the user to specify the `TripletFields` object for performance improvement as we explained previously. In addition to API improvements, `aggregateMessages` is optimized for performance.

Since `mapReduceTriplets` is now deprecated, we will not discuss it further. If you have to use it with earlier versions of Spark, you can refer to the Spark programming guide.

Summary

`AggregateMessages` provides a functional abstraction for aggregating neighborhood information in Spark graphs. This operator applies a user-defined `sendMsg` function to each edge in the graph using `EdgeContext`. Each `EdgeContext` class accesses the required information about the edge and passes that information to its source node and/or destination node using the `sendToSrc` and/or `sendToDst` methods respectively. After all messages have been received by the nodes, the `mergeMsg` function is used to aggregate those messages at each node.

In the next chapter, we will introduce another operator called `Pregel`, which will be useful for creating custom iterative graph-processing algorithms.

6
Iterative Graph-Parallel Processing with Pregel

Graphs are a very useful abstraction for solving many practical computing problems. For example, we can search through nearly five billion web pages today, thanks to the PageRank graph algorithm. Apart from the web search, there are other applications, such as social media, for which iterative graph processing is needed. In this chapter, we will learn how to use **Pregel**, a computational model, which is suitable for this task. Pregel was initially proposed by Google and has also been adopted by Spark as a generic programming interface for iterative graph computations. In this chapter, you will understand the Pregel model of computation. In addition, our learning goal is to clarify both the interface and implementation of the Pregel operator in Spark. After working through the concrete examples, you will be able to formulate your own algorithms with the Pregel interface.

The Pregel computational model

A Pregel program is a sequence of iterations called **supersteps**, in each of which a vertex can receive inbound messages that are sent by its neighbors in the previous iteration, and modify its attribute and its edges. In addition, each vertex also sends messages to its neighbors by the end of each superstep. By thinking as a vertex, this abstraction makes it simple to reason about parallel graph processing. All we need to think about is the type of message that each vertex should be receiving, the processing that it should do on its inbound messages, and the message that its neighbors need for the next superstep. Luckily, this message-passing approach is flexible enough to express a large class of graph algorithms. More importantly, a graph algorithm can make use of Spark's scalable architecture to process the messages in bulk and in a synchronous manner. This synchronous model of computation makes it easy to express most graph-parallel algorithms.

Example – iterating towards the social equality

Before presenting the Pregel API, let's illustrate these concepts with a hypothetical example of a social network, in which each person is extremely altruistic. We will assume that everyone knows how much money their friends have in the banks. However, they need an algorithm that will attempt to equalize their wealth. This is just an example (luckily or sadly, depending on your philosophy), but it will help clarify how Pregel works. In essence, each person will compare their money with their friends, and will send some of it to those who have less.

By using Pregel, they will equalize their wealth by sending money to each other through a sequence of iterations. In this case, we can use `Double` as the message type for our algorithm. In the beginning of each iteration, each person will first receive a sum of money that was donated by their friends in the previous iteration. Based on their knowledge of how much their friends now own, they will compare their new wealth against their friends' situations. This means they need to find out who earn less, and then calculate how much they should send to those friends. At the same time, they also decide how much to keep in their account. As we described it, each Pregel iteration consists of three consecutive tasks, and this is why it is called a *superstep*.

Hence, they first need a function called `mergeMsg` to combine the inbound money transfers that they may receive from their well off friends:

```
def mergeMsg(fromA: Double, fromB: Double): Double = fromA + fromB
```

Second, they will also need a function, called **vertex program**, to calculate how much money they have after receiving money in the previous superset:

```
def vprog(id: VertexId, balance: Double, credit: Double) = balance
+ credit
```

Finally, a function called `sendMsg` is also needed for sending money between friends:

```
def sendMsg(triplet: EdgeTriplet[VD, ED]): Iterator[(VertexId, A)]
```

As seen from the previous function signature, `sendMsg` takes an edge triplet as an input instead of a vertex, so that we have access to both the source and destination nodes. We figure out the correct implementation of `sendMsg` in the next section.

Let's further simplify our example by considering a triangle network between three friends:

```
scala> val nodes: RDD[(Long,Double)] = sc.parallelize(List((1,10.0),(2,3.0),(3,5.0)))
nodes: RDD[(Long, Double)]
```

```
scala> val edges = sc.parallelize(List(Edge(1,2,1),Edge(2,1,1),Edge(1,3,1
),Edge(3,1,1),Edge(2,3,1),Edge(3,2,1)))
edges: Edge[Int]]
scala> val graph = Graph(nodes, edges)
graph: Graph[Double,Int]

scala> graph.vertices.foreach(println)
(1,10.0)
(2,3.0)
(3,5.0)
```

For simplicity, assume that each person will distribute five percent of her wealth to each of its poor friends. She will not need to worry if a friend receives too much since selfishness and greed are out of the equation here. So, here is our first attempt at implementing the sendMsg function:

```
def sendMsg(t: EdgeTriplet[Double, Double]) =
    if (t.srcAttr <= t.dstAttr) Iterator.empty
    else Iterator((t.dstId,t.srcAttr * 0.05))
```

This seems reasonable. If a person is better off than her friends, she will offer five percent of her money to that friend. Otherwise, she gives nothing. After ten iterations, our new graph thus becomes:

```
val newGraph = graph.pregel(0.0,10)(vprog, sendMsg, mergeMsg)
```

> Notice that Pregel takes two argument lists (for example, graph.pregel(list1)(list2)). The first argument list includes an initial message to send to all vertices in the beginning of the algorithm as well as the maximum number of iterations. The second argument list contains the three user-defined functions for combining, receiving, and computing messages.

```
scala> newGraph.vertices.foreach(println)
(3,10.951096875000001)
(2,10.246937500000001)
(1,10.512346875)
```

Something is not right here. The group started with 18 dollars in total, and ended up with more than 30 dollars. This cannot be true! So, what did we do wrong? To uncover our mistake, let's see what happened after one iteration of Pregel:

```
val newGraph1 = graph.pregel(0.0,1)(vprog, sendMsg, mergeMsg)
```

```scala
scala> newGraph1.vertices.foreach(println)
(1,10.0)
(2,3.75)
(3,5.5)
```

Again, their total wealth exceeds 18 dollars after one iteration. This is because when a person sent an amount of money to their friend, that amount was not debited from that person's account. We can fix this by sending messages to the person that received the money as well as the one that sent it. So, if person *A* sends *X* dollars to person *B*, we should send *X* dollars to *B*, and -*X* dollars to *A*:

```scala
def sendMsg(t: EdgeTriplet[Double, Double]) =
    if (t.srcAttr <= t.dstAttr) Iterator.empty
    else  Iterator((t.dstId, t.srcAttr * 0.05),
                    (t.srcId, - t.srcAttr * 0.05))

val afterOneIter = graph.pregel(0.0, 1)(vprog, sendMsg, mergeMsg)
```

Let's see the output:

```scala
scala> afterOneIter.vertices.foreach(println)
(1,9.0)
(2,3.75)
(3,5.25)
```

You can verify that things now work as expected. So, what if we increase the maximum number of iterations? Let's see what happens then:

```scala
scala> afterTenIters.vertices.foreach(println)
(1,5.999611965064453)
(2,6.37018749852539)
(3,5.630200536410156)

scala> afterHundredIters.vertices.foreach(println)
(1,6.206716647163644)
(2,6.207038273723298)
(3,5.586245079113054)
```

Even with 100 iterations, we can see that the account balances do not converge to the idealistic value of 6 dollars, but fluctuate around it. This is expected in our simplistic example.

The Pregel API in GraphX

Now, let's formalize the programming interface for the Pregel operator. Here is its definition:

```
class GraphOps[VD, ED] {
  def pregel[A]
      (initialMsg: A,
       maxIter: Int = Int.MaxValue,
       activeDir: EdgeDirection = EdgeDirection.Out)
      (vprog: (VertexId, VD, A) => VD,
       sendMsg: EdgeTriplet[VD, ED] => Iterator[(VertexId, A)],
       mergeMsg: (A, A) => A)
    : Graph[VD, ED]
}
```

The `pregel` method is invoked on a property graph, and returns a new graph with the same type and structure. While the edges remain intact, the attributes of the vertices may change from one superset to the next one. Pregel takes the following two lists of arguments. The first list contains:

- An initial message with a user-defined type A — this message is received by each vertex when the algorithm starts

- A maximum number of iterations

- The edge direction along which to send messages

> A Pregel algorithm terminates when either there are no more messages to be sent, or when a specified maximum number of iterations is reached. When implementing an algorithm, it is important to always limit the number of iterations, especially when the algorithm is not guaranteed to converge.
>
> If no active edge direction is specified, Pregel assumes that messages are only sent for the outgoing edges of each vertex. Moreover, if a vertex did not receive a message in the previous superset, no message will be sent along its outgoing edge, at the end of the current superset.

In addition, the second list of arguments must include the three functions:

```
vprog: (VertexId, VD, A) => VD: this vertex program updates the
attributes of all vertices who received messages from the previous
iteration
mergeMsg: (A, A) => A: this function merges the messages to be
received by each vertex.
sendMsg: EdgeTriplet[VD, ED] => Iterator[(VertexId, A)]: this
function takes an edge triplet and creates the messages to be sent
to the source node and/or destination node.
```

Community detection through label propagation

In the following section, we are going to implement a community detection algorithm using the Pregel interface. **Label Propagation Algorithm (LPA)** is a simple and fast method for detecting communities within graphs. By construction, the communities obtained by the label propagation process require each node to have at least as many neighbors within its community as it has with each of the other communities.

Let's quickly describe how the LPA works. First, each node is initially given its vertex ID as its label. At the subsequent iterations, each node determines its community, based on the labels of its neighbors. Specifically, the node chooses to join the community to which the maximum number of its neighbors belong to. If there is a tie, one of the majority labels is picked randomly. As we propagate the labels in this way across the graph, most labels will disappear, whereas the remaining ones define the communities. Ideally, this iterative algorithm converges when no node in the network changes its label. As a result, nodes having the same labels are grouped together as one community.

By implementing this algorithm in Pregel, we want to obtain a graph in which the vertex attributes are the labels of the community affiliations. Hence, we'll first initialize the LPA graph by setting the label of each vertex to its identifier:

```
val lpaGraph = graph.mapVertices { case (vid, _) => vid }
```

Next, we'll define the type of message to `Map[Label, Long]`, which associates a community label to the number of neighbors that have this label. The initial message that will be sent to each node is simply an empty map:

```
type Label = VertexId
val initialMessage = Map[Label, Long]()
```

Following the Pregel programming model, we define a `sendMsg` function, which is used by each node to inform its neighbors of its current label. For each triplet, the source node will receive the destination node's label, and vice versa:

```
def sendMsg(e: EdgeTriplet[Label, ED]): Iterator[(VertexId,
Map[Label, Long])] =
    Iterator((e.srcId, Map(e.dstAttr -> 1L)), (e.dstId,
    Map(e.srcAttr -> 1L)))
```

After receiving the messages from its neighbors, a node determines its community label as the one to which the majority of its neighbors currently belong to. Hence, each node will use the following vertex program function to do so:

```
def vprog(vid: VertexId, attr: Long, message: Map[Label, Long]):
VertexId = if (message.isEmpty) attr else message.maxBy(_._2)._1
```

The previous function returns, in each iteration, the label (that is, a `VertexId` attribute) of the community to which the majority of its neighbors currently belong to.

We also need a `mergeMsg` function to combine all the messages, received by a node from its neighbors into a single map. If both the messages contain the same label, we simply sum up the corresponding number of neighbors for this label:

```
def mergeMsg(count1: Map[Label, Long], count2: Map[Label, Long])
  : Map[VertexId, Long] = {
  (count1.keySet ++ count2.keySet).map { i =>
    val count1Val = count1.getOrElse(i, 0L)
    val count2Val = count2.getOrElse(i, 0L)
    i -> (count1Val + count2Val)
  }.toMap
}
```

Finally, we can run the LPA algorithm as we did for equalizing the social wealth by calling the `pregel` method on the graph:

```
lpaGraph.pregel(initialMessage, 50)(vprog, sendMsg, mergeMsg)
```

> The main benefits of LPA are its simplicity and time efficiency. In fact, the number of iterations to convergence has been observed to be independent of the graph size whereas each iteration has a linear time complexity. Despite its advantages, the label propagation algorithm may not necessarily converge and it may also result in uninteresting solutions, such as each node being identified as a single community. Actually, the algorithm may oscillate for graphs that are bipartite or have a nearly bipartite structure.

The Pregel implementation of PageRank

We have already seen that GraphX has a PageRank API. In the following, let us see how this famous web search algorithmic can be easily implemented using Pregel. Since we already explained in the previous chapter how PageRank works, we will now simply explain its Pregel implementation:

First of all, we need to initialize the ranking graph with each edge attribute set to 1, divided by the out-degree, and each vertex attribute to set 1.0:

```
val rankGraph: Graph[(Double, Double), Double] =
    // Associate the degree with each vertex
    graph.outerJoinVertices(graph.outDegrees) {
        (vid, vdata, deg) => deg.getOrElse(0)
    }.mapTriplets( e => 1.0 / e.srcAttr )
     .mapVertices( (id, attr) => (0.0, 0.0) )
```

Following the Pregel abstraction, we define the three functions that are needed to implement PageRank in GraphX. First, we define the vertex program as follows:

```
val resetProb = 0.15
def vProg(id: VertexId, attr: (Double, Double), msgSum: Double):
(Double, Double) = {
  val (oldPR, lastDelta) = attr
  val newPR = oldPR + (1.0 - resetProb) * msgSum
  (newPR, newPR - oldPR)
}
```

Next is the function that creates the messages:

```
val tol = 0.001
def sendMessage(edge: EdgeTriplet[(Double, Double), Double]) = {
  if (edge.srcAttr._2 > tol) {
    Iterator((edge.dstId, edge.srcAttr._2 * edge.attr))
  } else {
    Iterator.empty
  }
}
```

The third function called `mergeMsg` simply adds the rank:

```
def mergeMsg(a: Double, b: Double): Double = a + b
```

Then we will get the vertex ranking as follows:

```
rankGraph.pregel(initialMessage, activeDirection =
EdgeDirection.Out)
                (vProg, sendMsg, mergeMsg)
                .mapVertices((vid, attr) => attr._1)
```

Summary

In summary, Pregel is a generic and simplified interface for writing custom iterative, and parallel algorithms on large graphs. In this chapter, we have seen how to implement different iterative graph processing using this simple abstraction. In the next chapter, we will see how to use Spark's MLlib and GraphX to solve some machine learning problems with graph data.

7

Learning Graph Structures

In this chapter, we will show you how to learn interesting structures from graphs in Spark. In principle, one learns and finds relationships from data by first selecting the problem of interest. The most common learning problems are regression, classification, ranking, and clustering. In this book, we will focus on clustering. In particular, we will focus on graph data, and apply clustering to detect communities within the graphs. Here is our roadmap for this chapter. First, we will introduce the concepts of spectral clustering. Then, we will study a specific method, which allows us to cluster graphs in Spark. Finally, we will apply these techniques to music and song playlist datasets. This application will also serve as an opportunity to review the tools and techniques that we covered in the previous chapters. We will bring them together in this chapter.

Community clustering in graphs

Clustering is a learning problem in which given entities, such as objects or people, are partitioned into subsets, according to a defined similarity measure. The entities within the same cluster are very similar, and are different from all entities in other clusters. Clustering is done with an unsupervised method. In other words, it operates on unlabeled data, which are the attributes or features of the entities. Moreover, clustering methods can be broadly classified into parametric versus non parametric approaches. The parametric approaches impose a probability model on the data. Some examples of the parametric methods are **Gaussian Mixture Model (GMM)** and **Latent Dirichlet Allocation (LDA)**. On the other hand, the non parametric models infer the structure of the clusters from the data itself. Examples include k-means and spectral clustering. All these cited methods are available in Spark's MLlib library.

Before we continue, it is important to understand why clustering is related to graph processing. There are two reasons for this. The first reason is that clustering is very useful for detecting "communities" in graphs. These communities are essentially clusters of nodes that share similar features. While two nodes are not explicitly connected, clustering can reveal their similarities by learning from their attribute data. For instance, online social and dating websites use such information to suggest people you may know, or the partners you would be interested to meet. Conversely, clustering can be helpful to uncover interesting structures in highly connected networks. The second reason is that the clustering method that we will see here is based on graph processing. In particular, we will focus our attention on the **power iteration clustering** (PIC), which is a simple and fast spectral clustering method.

Spectral clustering

As mentioned previously, the aim of clustering is to divide the data points into several clusters in such a way that the points in the same cluster are very similar, and the points in different clusters are dissimilar to each other. A "similarity graph" is a nice way to represent the similarities of the data points. Each point becomes a node in the similarity graph, whereas each edge has, as its attribute, the "similarity measure" of the connected nodes. As a result, the clustering problem reduces to find a partition of a graph in such a way that the edges between the different groups have very low weights, and the edges within a group have high weights. To do this, we use the spectral clustering technique, which basically reduces the high-dimensional similarity graph to a low-dimensional representation. To keep our discussion really simple, we will avoid the math. However, the technical details can be learned by reading some good tutorials, such as the one available at http://arxiv.org/abs/0711.0189.

Power iteration clustering

An efficient and scalable spectral clustering method is the power iteration clustering (PIC) method. It is defined in the MLlib library, precisely in http://spark.apache.org/docs/latest/mllib-clustering.html#power-iteration-clustering-pic. It is implemented in Spark using GraphX's processing APIs, and the caching optimizations. Here is the API for this PIC clustering method:

```
class PowerIterationClustering {

    // Run the PIC algorithm.
    def run(similarities: RDD[(Long, Long, Double)]):
    PowerIterationClusteringModel

    // Set the initialization mode. Either "random" or "degree"
```

```
def setInitializationMode(mode: String):
PowerIterationClustering.this.type

// Set the number of clusters.
def setK(k: Int): PowerIterationClustering.this.type

// Set maximum number of iterations of the power iteration
loop
def setMaxIterations(maxIterations: Int):
PowerIterationClustering.this.type
}
```

To apply the PIC clustering to graphs, we will need to follow these five steps:

1. First, load the data into a Spark graph property.
2. Second, extract the features of the nodes.
3. Third, define a similarity measure between the two nodes.
4. Next, create an affinity matrix, based on the initial graph using the similarity measure.
5. Finally, run the k-means clustering on the affinity matrix.

Steps 1 and 2 can be done using the graph builder methods that we learned in *Chapter 2, Building and Exploring Graphs*. Step 3 simply requires us to define a function that determines how similar the two nodes are. The choice of similarity measure depends on the nodes' features, and the problem at hand. Nonetheless, there exists standard measures from which we can choose. For instance, if the node feature is a binary vector, we can use the Jaccard similarity. On the other hand, a Gaussian kernel function can be used when the node feature is a real vector. These are not the only possibilities, and we can also define our own measure.

In Step 4, the affinity matrix `similarities` should be represented by an RDD of (i, j, sim) tuples. The similarity *sim* must be a nonnegative number. For any edge $(i\,j)$ with a nonzero similarity, there should be either (i, j, sim), or (j, i, sim) in the input. Since the affinity matrix must be symmetric, if only (i, j, sim) is available in the data, the reciprocal (j, i, sim) is assumed, and vice versa. Moreover, tuples with $i = j$ are simply ignored.

The last step consists of two steps. First, we create `PowerIterationClusteringModel` from the `similarities` matrix, and then we run a k-means clustering on it. Before running the clustering model, we must also choose two parameters:

*   The maximum number of iterations for the k-means clustering
*   The maximum number of clusters, K

A sketch of the application of PIC is shown in the following code:

```
import org.apache.spark.mllib.clustering.PowerIterationClustering

// Define pairwise similarities based on initial graph
val similarities: RDD[(Long, Long, Double)] = ...

// Create the PIC clustering model
val pic = new PowerIteartionClustering()
  .setK(maxClusterNumber)
  .setMaxIterations(maxIterations)

// Run the PIC clustering model
val clusteringResult: RDD[Assignment] =
pic.run(similarities).assignments

clusteringResult.collect().foreach { a =>
  println(s"${a.id} -> ${a.cluster}")
}
```

The PIC method returns an RDD of assignment, which abstracts a tuple of `VertexId`, and `Int` that corresponds to the node ID, and its cluster group.

Applications – music fan community detection

We are now ready to apply the previous graph clustering method to the cluster music songs, according to the tags attached to each song. Alternatively, a dataset of the song playlists can also be used to cluster songs that are often played in many lists. The datasets that we are going to work with can be downloaded from `http://www.cs.cornell.edu/~shuochen/lme/data_page.html`. The datasets consist of the following files:

- `train.txt`: This file contains the playlist data by using the integer ID to represent songs
- `tags.txt`: This file includes the social tags by using the integer ID to represent songs
- `song_hash.txt`: This file maps a song ID to its title and artist
- `tag_hash.txt`: This one maps a tag ID to its name

Each file has a particular format as explained here:

- **Format of the playlist data**: The first line of the data file consists of the IDs (not the integer ID, but the IDs from other sources for identifying the songs) for the songs, separated by a space. We will not need this first line here, and thus it can be skipped. The second line consists of the number of appearances of each song in the file, also separated by a space. Starting from the third line are the playlists, with each song represented by its integer ID in this file (from 0 to the total number of songs minus one). Note that in the playlist data file, each line ends with a space.

- **Format of the tag data**: The tag data file has the same number of lines as the total number of songs in the playlist data. Each line consists of the IDs of the tags for a song, represented by integers, and separated by space. If a song does not have a tag, its line is just a #. Note that for the tag file, there is no space at the end of each line.

- **Format of the song mapping file**: Each line corresponds to one song, and has the format called `Integer_ID\tTitle\tArtist\n`.

- **Format of the tag mapping file**: Each line corresponds to one song, and has the format called `Integer_ID, Name\n`.

First, let's follow the previous five steps to cluster the songs by their tags.

Step 1 – load the data into a Spark graph property

We define a class song. Each song has as its attributes a title, an artist name, and a set of tags:

```scala
scala> case class Song(title: String, artist: String, tags: Set[String])
{
        override def toString: String = title + ", "  + artist
    }
defined class Song
```

Now, we import the songs into `RDD[(VertexId, Song)]`, and initialize each song with an empty set of tags:

```scala
scala> var songs: RDD[(VertexId, Song)] =
        sc.textFile("./data/song_hash.txt").
        map {line =>
            val row = line split '\t'
```

```
            val vid = row(0).toLong
            val song =  Song(row(1), row(2), Set.empty)
            (vid, song)
        }
songs: RDD[(VertexId, Song)]
```

Then, we can create a graph property, whose nodes are the songs. It will not add any edges into the graph at first, and will simply pass an empty RDD to Graph.apply:

```
scala> val graphFromSongs: Graph[Song, Int] = {
            val zeroEdge: RDD[Edge[Int]] = sc.parallelize(Nil)
            Graph(songs, zeroEdge)
        }
graphFromSongs: Graph[Song,Int]


scala> graphFromSongs.vertices.take(5).foreach(println)
(1084,Song(Tequila Sunrise,Fiji,Set()))
(1410,Song(The Sweetest Taboo,Sade,Set()))
(3066,Song(Bow Chicka Wow Wow,Mike Posner,Set()))
(1894,Song(Love Your Love The Most,Eric Church,Set()))
(466,Song(Stupify,Disturbed,Set()))
```

Step 2 – extract the features of nodes

Now, let's join the tags from the dataset called tags.txt into the nodes. To do this, we first need to create RDD[(VertexId, Set[String])], which we will then join into graphFromSong:

```
scala> val tagIter: Iterator[(VertexId, Set[String])] =
            Source.fromFile("./data/tags.txt").getLines.zipWithIndex.
            map {
                x =>
                val tags = x._1 split ' '
                (x._2.toLong, tags.toSet)
            }
tagIter: Iterator[(VertexId, Set[String])] = non-empty iterator


scala> val tagRDD = sc.parallelize(tagIter.toSeq)
tagRDD: RDD[(VertexId, Set[String])]
```

For now, we have only the mapping between the song ID and the set of tag IDs in our `tagRDD`:

```scala
scala> tagRDD.take(3).foreach(println)
(0,Set(115, 173))
(1,Set(62, 88, 110, 90, 123, 155, 173, 14, 190, 214, 115, 27))
(2,Set(115, 173))
```

What we want is to extract the tag names from `tag_hash.txt` given the tag ID. We can now call `joinVertices` on `graphFromSongs`, and pass the RDD of tags `tagRDD` with a function that extracts the tags. Note that in the dataset called `tags.txt`, a # tag assigned next to the song ID means that no tag is associated with that song. In such a case, we simply return the initial song with an empty tag. Otherwise, we add the set of tags into the song:

```scala
scala> val songsNtags = graphFromSongs.joinVertices(tagRDD){
        (id, s, ks) => ks.toList match {
           case List("#") => s
           case _         => {
               val tags: Map[Int, String] =
               Source.fromFile("./data/tag_hash.txt").getLines().
               map {
                   line =>
                   val row  = line split ", "
                   row(0).toInt -> row(1)
               }.toMap

               val songTags = ks.map(_.toInt) flatMap (tags get)
               Song(s.title, s.artist, songTags.toSet)
           }
        }
      }
songsNtags: Graph[Song,Int]

scala> songsNtags.vertices.take(3).foreach(println)
(1084,Tequila Sunrise, Fiji)
(1410,The Sweetest Taboo, Sade)
(3066,Bow Chicka Wow Wow, Mike Posner)
```

Step 3 – define a similarity measure between two nodes

Since we want to cluster the songs by their social tags, a natural way to measure the similarity between two songs is the Jaccard metric. Simply put, it is the ratio of the number of common tags between two songs, and their total number of tags. If none of the songs is tagged, we assume that their similarity score is zero:

```
def similarity(one: Song, other: Song):Double = {
        val numCommonTags = (one.tags intersect other.tags).size
        val numTotalTags = (one.tags union other.tags).size
        if (numTotalTags > 0)
            numCommonTags.toDouble / numTotalTags.toDouble
        else 0.0
}
```

Step 4 – create an affinity matrix

Now, we need to calculate the similarity between each pair of songs in our database. If there are 1,000 songs, we will have to compute, and store, one million similarity scores. What if we had 1,000,000 songs? Obviously, computing similarities between every pair will be inefficient. Instead, we can restrict this to the songs that have a relatively high similarity score. At the end of the day, we want to cluster songs that are similar. Therefore, we will filter the nodes with the following function:

```
scala> def quiteSimilar(one: Song, other: Song, threshold: Double):
Boolean = {
        val commonTags = one.tags intersect other.tags
        val combinedTags = one.tags union other.tags
        commonTags.size > combinedTags.size * threshold
    }
quiteSimilar: (one: Song, other: Song, threshold: Double)Boolean
```

This next function helps to remove the duplicate songs in our graph data:

```
def differentSong(one: Song, other: Song): Boolean =
    one.title != other.title || one.artist != other.artist
```

With these two functions, we can now create RDD[Edge[Double]] that will contain a similarity measure between the nodes that are quite similar:

```
// First, get the songs with tags
songs = songsNtags.vertices

// Then, compute the similarity between each pair of songs
```

```
// with a similarity score larger than 0.7
val similarConnections: RDD[Edge[Double]] = {
    val ss = songs cartesian songs
    val similarSongs = ss filter {
        p => p._1._1 != p._2._1 &&
        similarByTags(p._1._2, p._2._2, 0.7) &&
        differentSong(p._1._2, p._2._2)
    }

    similarSongs map {
        p => {
            val jacIdx = similarity(p._1._2, p._2._2)
            Edge(p._1._1, p._2._1, jacIdx)
        }
    }
}
```

A simple check shows that we only need to store 1,506 similarity scores instead of 10 million:

```
scala> similarConnections.count
res8: Long = 1506
scala> songs.count
res9: Long = 3168

scala> 3168 * 3168
res10: Int = 10036224
```

While we are at it, let's create our similarity graph:

```
scala> val similarByTagsGraph = Graph(songs, similarConnections)
```

Some of our songs have very few tags, so let's filter those out:

```
val similarHighTagsGraph = similarByTagsGraph.subgraph(vpred =
(id: VertexId, attr: Song) => attr.tags.size > 5)
```

Let's check the output:

```
scala> similarHighTagsGraph.vertices.count
res12: Long = 2144
scala> similarHighTagsGraph.edges.count
res13: Long = 126
```

Let's look closer into the graph:

```
scala> similarHighTagsGraph.triplets.take(6).foreach(t => println(t.
srcAttr + " ~~~ " + t.dstAttr + " => " + t.attr))
Fancy (w\/ T.I. & Swizz Beatz), Drake ~~~ Any Girl (w\/ Lloyd), Lloyd
Banks => 0.8571428571428571
Roll With It, Easton Corbin ~~~ You Lie, The Band Perry =>
0.7142857142857143
Any Girl (w\/ Lloyd), Lloyd Banks ~~~ Fancy (w\/ T.I. & Swizz Beatz),
Drake => 0.8571428571428571
Any Girl (w\/ Lloyd), Lloyd Banks ~~~ I'm Going In (w\/ Young Jeezy & Lil
Wayne), Drake => 0.7142857142857143
Everything Falls, Fee ~~~ Needful Hands, Jars Of Clay =>
0.7142857142857143
Bring The Rain, MercyMe ~~~ Needful Hands, Jars Of Clay => 0.75
```

So, we can see that `Fancy` by `Drake` is similar to `Any Girl` by `Lloyd Banks`. Of course, they are rap songs.

Let's finally create the affinity matrix of type `RDD[(Long, Long, Double)]`, which is needed to run the PIC algorithm:

```
val similarities: RDD[(Long,Long,Double)] =
    similarHighTagsGraph.triplets.map{t => (t.srcId, t.dstId, t.attr)}
```

Step 5 – run k-means clustering on the affinity matrix

We can choose the number of clusters to be $K = 7$:

```
scala> val similarities: RDD[(Long,Long,Double)] = similarHighTagsGraph.
triplets.map{t => (t.srcId, t.dstId, t.attr)}
scala> val pic = new PowerIterationClustering().setK(15).
setMaxIterations(20)
pic: org.apache.spark.mllib.clustering.PowerIterationClustering


scala> val clusteringModel = pic.run(similarities)
clusteringModel: org.apache.spark.mllib.clustering.
PowerIterationClusteringModel


scala> clusteringModel.assignments.foreach { a =>
     |      println(s"${a.id} -> ${a.cluster}")
     | }
327 -> 0
```

```
715 -> 0
3063 -> 2
2879 -> 2
1623 -> 0
3003 -> 0
2539 -> 0
2283 -> 0
2163 -> 0
2979 -> 0
2615 -> 5
2147 -> 1
2667 -> 3
2531 -> 0
2149 -> 4
```

Extra Step: Looking into the clustering results

Well, we cannot really see anything through these numbers. So, let's explore the clusters and see what common tags the songs within each cluster have.

First, let's use the results of the clustering to create a graph whose nodes contain the actual song, as well as its clustering ID. To do this, we use the VertexRDD collection's innerJoin method twice. First, we join the clustering assignment to the graph of songs. Since innerJoin can alter the attribute type of the vertices, it does not matter what the initial graph's vertex type is. For simplicity, we initialize each vertex attribute to 0.0. The second application of innerJoin joins the VertexRDD collection of songs into the result of the first application:

```
val clustering: RDD[(Long, Int)] =
clusteringModel.assignments.map(a => (a.id, a.cluster))

val graph: VertexRDD[Double] =
Graph.fromEdges[Double,Double](similarities.map(t =>
Edge(t._1,t._2,t._3)), 0.0).vertices

val clusteredSongs: VertexRDD[(Song, Int)] =
graph.innerJoin(clustering){ (id, _, cluster) => cluster
}.innerJoin(songs){ (id, cluster, s) => (s, cluster)}
```

As a result, we obtain a new VertexRDD collection clusteredSongs, which contains both the songs and their cluster IDs:

```
scala> clusteredSongs.first
res25: (VertexId, (Song, Int)) = (2372,(Hold My Heart, Tenth Avenue
North,7))
```

We can put this into a property graph with the similarity scores:

```
scala> val clusterNScoreGraph = Graph(clusteredSongs, similarities.map(t
=> Edge(t._1,t._2,t._3)))
clusterNScoreGraph: Graph[(Song, Int),Double]
```

```
scala> clusterNScoreGraph.triplets.first
res37: EdgeTriplet[(Song, Int),Double] = ((38,(Fancy (w\/ T.I.
& Swizz Beatz), Drake,2)),(1976,(Any Girl (w\/ Lloyd), Lloyd
Banks,2)),0.8571428571428571)
```

Because not all the neighboring songs correspond to the same cluster, we can filter out the edges between two songs that belong to two different clusters. In other words, these songs have some similarity, but they do not really belong together:

```
scala> val clusteredSongGraph = clusterNScoreGraph.subgraph(epred = t =>
t.srcAttr._2 == t.dstAttr._2)
clusteredSongGraph: Graph[(Song, Int),Double]
```

```
scala> clusteredSongGraph.edges.count
res5: Long = 50
```

Next, we replace the attribute of the remaining edges by the set of common tags between the songs that they connect. This can be easily done, thanks to the `mapTriplets` method:

```
val clusteredTagsGraph = clusteredSongGraph.mapTriplets(t =>
t.srcAttr._1.tags intersect t.dstAttr._1.tags)
```

Let's see what we get:

```
scala> clusteredTagsGraph.triplets.take(3).foreach(println)
((482,(Roll With It, Easton Corbin,0)),(2866,(You Lie, The Band
Perry,0)),Set(new country, modern country, country, great song, my
favorite))
((1976,(Any Girl (w\/ Lloyd), Lloyd Banks,6)),(2470,(I'm Going In (w\/
Young Jeezy & Lil Wayne), Drake,6)),Set(rap, wdzh-fm, wjlb-fm, whtd-fm,
wkqi-fm))
((2364,(While I'm Waiting, John Waller,0)),(2372,(Hold My Heart,
Tenth Avenue North,0)),Set(worship, favorite, christian, contemporary
christian, christian rock))
```

Now, if we want to approximately find the common tags in each cluster, we can make use of the Pregel operator to do so. Remember that the Pregel implementation in Spark allows the passing of the message only between neighboring nodes. However, our `clusteredTagsGraph` has nodes that are not directly connected by an edge, but still belong to the same cluster, maybe through other nodes. Thus, the Pregel operator will not find the absolute intersection of tags in each cluster, but it will still be helpful to see some patterns:

```
val commonTagsByCluster =
clusteredTagsGraph.pregel[Set[String]](initialMsg = Set.empty,
maxIterations = 10){
    (id, sc, m) => sc,
    t => Iterator((t.srcId, t.srcAttr._1.tags intersect
    t.dstAttr._1.tags),
    (t.dstId, t.srcAttr._1.tags intersect t.dstAttr._1.tags)),
    (s1, s2) => s1 intersect s2
}
```

Looking at the results, we can find some straightforward clustering. The cluster #1 consists of worship songs:

```
scala> commonTagsByCluster.triplets.filter(_.srcAttr._2 == 1).foreach(t
=> println(t.srcAttr._1 + " => " + t.attr))
Lead Me, Sanctus Real => Set(rock, worship, christian, contemporary
christian, christian rock, happy)
Needful Hands, Jars Of Clay => Set(rock, worship, christian, contemporary
christian, christian rock, favorites)
Revelation, Third Day => Set(rock, worship, christian, contemporary
christian, christian rock, favorites)
Give You Glory, Jeremy Camp => Set(rock, worship, christian, contemporary
christian, christian rock, favorites)
Revelation, Third Day => Set(rock, worship, christian, contemporary
christian, christian rock, favorites)
Cry Holy, Sonicflood => Set(rock, worship, christian, contemporary
christian, christian rock)
```

The cluster #2 is about country music and RnB songs:

```
scala> commonTagsByCluster.triplets.filter(_.srcAttr._2 == 2).foreach(t
=> println(t.srcAttr._1 + " => " + t.attr))
This Is Country Music, Brad Paisley => Set(beautiful, new country,
memories, country, great song, my favorite)
Bring It Back, Travis Porter => Set(wdzh-fm, hip hop, 2010s, wjlb-fm,
whtd-fm, energetic)
Anything Like Me, Brad Paisley => Set(beautiful, new country, memories,
country, great song, my favorite)
```

```
The Boys Of Fall, Kenny Chesney => Set(beautiful, new country, memories,
country, great song, my favorite)

Anything Like Me, Brad Paisley => Set(beautiful, new country, memories,
country, great song, my favorite)

Grove St. Party (w\/ Kebo Gotti), Waka Flocka Flame => Set(wdzh-fm, hip
hop, 2010s, wjlb-fm, whtd-fm, energetic, wkqi-fm)

The Boys Of Fall, Kenny Chesney => Set(beautiful, new country, memories,
country, great song, my favorite)

Make It Rain, Travis Porter => Set(wdzh-fm, hip hop, 2010s, wjlb-fm,
whtd-fm, energetic)

Grove St. Party (w\/ Kebo Gotti), Waka Flocka Flame => Set(wdzh-fm, hip
hop, 2010s, wjlb-fm, whtd-fm, energetic)

Love Faces, Trey Songz => Set(male vocalists, r&b, 2010s, wjlb-fm, rnb,
whtd-fm)

Words, Bobby V => Set(male vocalists, r&b, 2010s, wjlb-fm, rnb, whtd-fm)

Cupid, Lloyd => Set(male vocalists, r&b, 2010s, wjlb-fm, rnb, whtd-fm)
```

If you look at the other clusters, cluster #6 is hip hop. However, the last cluster #0 is less straightforward to tell. It is a mix and match of everything. Such a limitation is simply due to the imperfection of the social tag data, and not due to the PIC clustering method itself.

Exercise – collaborative clustering through playlists

Clustering music songs by social tags is not very effective. Imagine yourself having to tag every single song that you listen to. Instead of using explicit features, such as tags, we can alternatively use shared playlists to infer about the clustering. A playlist is a natural and more pervasive way to organize music. Now, the idea is that if two songs repeatedly appear in many lists, they are more likely to be similar than two other songs that do not belong to any cluster. I will leave the rest as an exercise. Just follow the same 5 steps that we took previously.

Summary

In this chapter, we studied how to solve the clustering problem for large-scale graphs. To do this, we introduced the Power Iteration Clustering method, and showed how to apply it to the clustering of songs using social tags. Using the song clustering example, we also reviewed the main graph building and processing techniques that we learned throughout this book. You should now be well-acquainted with using Spark's graph processing power to solve more interesting problems.

References

Chapter 2, Building and Exploring Graphs

Ahn,Y.-Y.; Ahnert, S.; Bagrow, J. P.; and Barabási, A.-L. *Flavor network and the principles of food pairing*, Nature, Scientific Reports 1, 196 (2011).

Klimmt, B. and Yang, Y. *Introducing the Enron corpus.* CEAS conference, 2004.

Leskovec, J.; Lang, K.; Dasgupta, A.; and Mahoney, M. *Community Structure in Large Networks: Natural Cluster Sizes and the Absence of Large Well-Defined Clusters.* Internet Mathematics 6(1) 29--123, 2009.

McAuley, J. and Leskovec, J. *Learning to Discover Social Circles in Ego Networks.* NIPS, 2012.

Page, L.; Brin, S.; Motwani, R., and Winograd, T. *The PageRank Citation Ranking: Bringing Order to the Web.* Technical Report. Stanford InfoLab, 1999.

Chapter 3, Graph Analysis and Visualization

Albert, R.; Jeong, H.; and Barabasi, A.-L. *Diameter of the World-Wide Web*. Nature, 1999.

Dutot, A.; Guinand, F.; Olivier, D.; and Pigné, Y. *GraphStream: A tool for bridging the gap between complex systems and dynamic graphs* in Emergent Properties in Natural and Artificial Complex Systems (EPNACS'07), Workshop of the 4th European Conference on Complex Systems (ECCS'07), Dresden, Germany.

Suereth, J. and Farewell, M. *SBT in Action: The simple Scala build tool*, Manning Publications, 2015.

Chapter 7, Learning Graph Structures

Lin, F. and Cohen W. W. (2010) *Power Iteration Clustering*, in Johannes Fürnkranz and Thorsten Joachims, ed., ICML, Omnipress, pp. 655-662.

Luxburg, U. von. *A Tutorial on Spectral Clustering*. Statistics and Computing 17(4): 395-416, 2007.

Index

A

aggregateMessages operator
  about 83
  aggregation, abstracting 85
  arguments, adding 88-90
  average point per game, calculating 90
  defense stats 91
  DRY principle 86-88
  EdgeContext parameter 83
analysis, of network connectedness
  about 43, 44
  clustering coefficients, computing 46-48
  connected components, finding 45, 46
  triangle, counting 46-48
Apache Zeppelin 36
average stats
  joining, into graph 92-95

B

bipartite food network example 75-78
bipartite graph
  building 22-26
BreezeViz
  about 36
  installing 36
  URL, for downloading 36

C

clustering 107
ColorBrewer
  URL 43
communication network 18

community clustering, in graphs
  about 107
  power iteration clustering (PIC) 108-110
  spectral clustering 108
community detection
  through label propagation 104, 105
compound nodes 18

D

degree distribution
  plotting, of network 41, 42
degree histogram, of social ego networks
  computing 32
degrees, in bipartite food network
  computing 31
degrees of network nodes
  computing 30
dependencies 52
directed graphs
  building 21, 22

E

edge attributes
  transforming 59, 60
EdgeContext parameter 83
edge directions
  reversing 72
edgeListFile graph builder 20
EdgeRDD
  data operations on 69
  joining 72
  mapping 69, 70
email communication graph 18

k-means clustering, running on affinity
matrix 116-120
similarity measure, defining between two
nodes 114

N

NCAA College Basketball datasets 79-82
neighboring information
collecting 74
network centrality 49
network connectedness
analysis 43-45
network datasets
about 17, 36
communication network 18
flavor network 18
social ego networks 19

O

out-degree, of Enron email network
computing 30, 31
outerJoinVertices operator 64

P

PageRank
about 49
working 49, 50
performance optimization 95-98
power iteration clustering (PIC)
about 108-110
reference link 108
Pregel 99
Pregel API, in GraphX 103
Pregel computational model
about 99
iterating, towards social equality 100-102
Pregel implementation,
of PageRank 105, 106
property graph 6

R

Resilient Distributed Dataset (RDD) 1
resolvers 52
reverse operator 62

S

Scala Build Tool
about 51
build definitions, organizing 51
library dependencies, managing 52
Scala Build Tool (SBT)
about 1, 14, 52
tasks, running with commands 58
URL 14
used, for building program 14, 15
share circle feature 19
social ego networks
about 19
reference link 19
Spark
URL, for documentation 46
URL, for download 2
Spark 1.4.1
downloading 1-3
installing 1-3
Spark application, building steps
about 53
build.sbt file, creating 53
library dependencies, declaring 54, 55
resolvers, declaring 54, 55
sbt-assembly plugin, enabling 53
sbt-assembly plugin, setting up 56
uber JAR, creating 57
Spark program
configuring 10-12
spark.app.name property 12, 13
spark.driver.memory property 13
spark.executor.memory property 13
spark.serializer property 13
spark.storage.memoryFraction
property 13
URL 12
writing 10-12
Spark shell
experimenting with 3, 4
spark-submit
used, for deploying standalone
application 15, 16
used, for running standalone
application 15, 16

Thank you for buying
Apache Spark Graph Processing

About Packt Publishing

Packt, pronounced 'packed', published its first book, *Mastering phpMyAdmin for Effective MySQL Management*, in April 2004, and subsequently continued to specialize in publishing highly focused books on specific technologies and solutions.

Our books and publications share the experiences of your fellow IT professionals in adapting and customizing today's systems, applications, and frameworks. Our solution-based books give you the knowledge and power to customize the software and technologies you're using to get the job done. Packt books are more specific and less general than the IT books you have seen in the past. Our unique business model allows us to bring you more focused information, giving you more of what you need to know, and less of what you don't.

Packt is a modern yet unique publishing company that focuses on producing quality, cutting-edge books for communities of developers, administrators, and newbies alike. For more information, please visit our website at www.packtpub.com.

About Packt Open Source

In 2010, Packt launched two new brands, Packt Open Source and Packt Enterprise, in order to continue its focus on specialization. This book is part of the Packt Open Source brand, home to books published on software built around open source licenses, and offering information to anybody from advanced developers to budding web designers. The Open Source brand also runs Packt's Open Source Royalty Scheme, by which Packt gives a royalty to each open source project about whose software a book is sold.

Writing for Packt

We welcome all inquiries from people who are interested in authoring. Book proposals should be sent to author@packtpub.com. If your book idea is still at an early stage and you would like to discuss it first before writing a formal book proposal, then please contact us; one of our commissioning editors will get in touch with you.

We're not just looking for published authors; if you have strong technical skills but no writing experience, our experienced editors can help you develop a writing career, or simply get some additional reward for your expertise.

[PACKT] open source
PUBLISHING   community experience distilled

Mastering Apache Maven 3

ISBN: 978-1-78398-386-5   Paperback: 298 pages

Enhance developer productivity and address exact
enterprise build requirements by extending Maven

1. Develop and manage large, complex projects
   with confidence.

2. Extend the default behavior of Maven with
   custom plugins, lifecycles, and archetypes.

3. Explore the internals of Maven to arm yourself
   with knowledge to troubleshoot build issues.

Apache Karaf Cookbook

ISBN: 978-1-78398-508-1   Paperback: 260 pages

Over 60 recipes to help you get the most out of your
Apache Karaf deployments

1. Leverage Apache Karaf to apply OSGi's
   powerful features to frameworks such as
   Apache ActiveMQ, Camel, Cassandra, CXF,
   and Hadoop.

2. Set up Apache Karaf for high availability.

3. A thorough guide with example-based recipes
   to help you get a deeper understanding of
   Apache Karaf's capabilities.

Please check **www.PacktPub.com** for information on our titles

www.ingramcontent.com/pod-product-compliance
Lightning Source LLC
Chambersburg PA
CBHW060147060326
40690CB00018B/4018